Enhanced Oil Recovery - Selected Topics

Edited by Badie I. Morsi
and Hseen O. Baled

Published in London, United Kingdom

Enhanced Oil Recovery - Selected Topics
http://dx.doi.org/10.5772/intechopen.97973
Edited by Badie I. Morsi and Hseen O. Baled

Contributors
Eman Mohamed Ibrahim Mansour, Nasser Mohammed Al Hinai, Ali Saeedi, Anastasia Ivanova, Alexey Cheremisin, Banwari Lal, Meeta Lavania, Nimmi Singh, Neha Sharma, Julio Gonzalo A. Herbas Pizarro, Afeez Gbadamosi, Adeyinka Yusuff, Augustine Agi, Jeffrey Oseh, Yani Carolina Araujo, Mariela Araujo, Badie I. I. Morsi, Hseen O. Baled, Asmaa Mohamed, Ahmed Mohamed Al Sabagh

First published in London, United Kingdom, 2022 by IntechOpen
IntechOpen is the global imprint of INTECHOPEN LIMITED, registered in England and Wales, registration number: 11086078, 5 Princes Gate Court, London, SW7 2QJ, United Kingdom

British Library Cataloguing-in-Publication Data
A catalogue record for this book is available from the British Library

Additional hard and PDF copies can be obtained from orders@intechopen.com

Enhanced Oil Recovery - Selected Topics
Edited by Badie I. Morsi and Hseen O. Baled
p. cm.
Print ISBN 978-1-80355-825-7
Online ISBN 978-1-80355-826-4
eBook (PDF) ISBN 978-1-80355-827-1

Meet the editors

Professor Morsi is the director of the Petroleum Engineering Program at the Swanson School of Engineering, University of Pittsburgh, USA. His research activities focus on reactor and process engineering; CO_2 capture from flue gas, fuel gas, air, and natural gas using chemical and physical solvents; enhanced oil recovery using CO_2 and alcohols; and CO_2 sequestration in depleted gas/oil reservoirs, deep coal seams, and underground saline aquifers. Professor Morsi received a BS (1972) in Petroleum Engineering from Cairo University, Egypt, and an MS (1977), a Ph.D. (1979), and an ScD (1982), all in Chemical Engineering from Ecole Nationale Supérieure des Industries Chimiques, University of Lorrain, France.

Dr. Hseen Baled is currently an Assistant Professor of Chemical and Petroleum Engineering at the University of Pittsburgh, USA, where he received his Ph.D. in Chemical Engineering. He was previously a research fellow at the National Energy Technology Laboratory, Pittsburgh. He graduated with a degree in Process Engineering from Rheinisch-Westfälishce Technische Hochschule (RWTH) Aachen University, Germany, in 20003. His research interests include the thermophysical properties of hydrocarbons at high pressure, high-temperature conditions, and technical challenges associated with oil recovery from ultra-deep petroleum reservoirs.

Contents

Preface

Oil production using enhanced oil recovery (EOR) methods has been strongly correlated with economics and crude oil prices; however, the general trend of the EOR market has been increasing despite the tumultuous economic and political environments over the past decades. We believe that the EOR market will continue to grow and play a key role in meeting the growing worldwide energy demand in the years to come despite the current environmental concerns and the ongoing transition to renewables, such as solar energy, biomass, and wind energy. We also believe that in the foreseeable future, renewables and nuclear energy alone will not be enough to replace the other primary energy sources and sustain or meet the growing energy demand of the steadily increasing world population.

Enhanced Oil Recovery - Selected Topics consists of nine chapters organized into five sections. Chapters address important topics on miscible flooding, surfactants flooding, polymers flooding, and microbial flooding.

- Chapter 1: "Introductory Chapter: Enhanced Oil Recovery"
- Chapter 2: "Carbon Dioxide-Oil Minimum Miscibility Pressure Methods Overview"
- Chapter 3: "Miscible Displacement Oil Recovery"
- Chapter 4: "Minimum Miscibility Pressure, Miscible Displacement, CO_2 Capture, and Injection"
- Chapter 5: "An Overview of Natural Surfactant Application for Enhanced Oil Recovery"
- Chapter 6: "Improving the Heavy Oil Recovery by Surfactants from Wastes"
- Chapter 7: "Chemical Enhanced Oil Recovery: Where Do We Stand? Recent Advances and Applications"
- Chapter 8: "Rock Lithology-Based Laboratory Protocols and Best Practices for Polymer Screening for EOR Application in Conventional and Harsh Reservoirs"
- Chapter 9: "Microbial Enhanced Oil Recovery: An Overview and Case Studies"

As editors, we diligently tried our best to maintain high-quality and plagiarism-free chapters, however, the authors of each chapter are ultimately responsible for the integrity and correctness of the materials presented in their respective chapters.

Badie I. Morsi and Hseen O. Baled
Department of Chemical and Petroleum Engineering,
Swanson School of Engineering,
University of Pittsburgh,
Pittsburgh, USA

Section 1

Introduction

Chapter 1

Introductory Chapter: Enhanced Oil Recovery

Hseen O. Baled and Badie I. Morsi

1. Introduction

The U.S. Energy Information Administration (EIA) predicts that the global energy consumption will increase by nearly 50% between 2020 and 2050 [1]. This is due mainly to the required industrial development to meet the rapid increase of the world's population, which is projected to reach about 9.7 billion by 2050, an increase of approximately 1.8 billion between now and then [2]. In 2020, the crude oil was the most-used energy source, accounting for more than 31% of the world's total primary energy consumption, which includes (oil, natural gas, coal, hydroelectric, nuclear energy, and renewables) as given in **Table 1** [3]. In 2021, the crude oil global demand is estimated to reach 108.2 million barrels per day (bpd) in 2045, which represents an increase of 17.6 million bpd between 2020 and 2045 [4]. Therefore, it is expected that crude oil production will continue to grow to support the predicted increase in energy consumption. Crude oil reserves, however, are depleting, and new discoveries of easy-to-find crude oil fields are becoming more difficult. Hence, improving oil recovery from the existing and depleted mature petroleum reservoirs through advanced recovery techniques has become a vital scheme in the oil and gas industry.

During the primary (natural drive and artificial lift) and secondary (waterflooding and immiscible gas injection) oil recovery stages, roughly one-third of the original oil-in-place can be recovered from mature reservoirs, implying that significant amounts of crude oil are left unrecovered following these two recovery stages. The remaining oil in the reservoirs is attributed to the low oil mobility and poor displacement efficiency, directly related to the high oil viscosity and surface tension, as well

Primary energy	Billion bbl (oil equivalent)	Share of total
Oil	28.47	31.3%
Natural Gas	22.49	24.7%
Coal	24.75	27.2%
Nuclear energy	3.92	4.3%
Hydroelectric	6.24	6.9%
Renewables	5.18	5.7%
Total	91.04	100.0%

(1 bbl of crude oil = 5.8 million BTU).

Table 1.
World primary energy consumption by fuel in 2020 [3].

as unfavorable interfacial fluid/fluid and fluid/rock properties. Therefore, tertiary recovery or enhanced oil recovery (EOR) methods have been thoroughly investigated and implemented to boost the production of the immobile oil left in the reservoirs by improving the microscopic and macroscopic (volumetric) sweep and oil displacement efficiencies.

EOR methods can be broadly classified into four major groups: thermal, chemical, miscible, and other methods. The thermal recovery methods primarily apply thermal energy (sensible heat) to heavy and highly viscous oil reservoirs as well as tar sands to reduce the oil viscosity, thus increasing its mobility. Examples of thermal recovery methods are steam injection, in-situ combustion (fire-flooding), and cyclic steam stimulation (huff-and-puff). Thermal recovery methods are well-established and have proven worldwide as successful EOR techniques. Among all thermal EOR processes, the steam injection has been remarkably successful and more commonly used due to its better control and improved heat transfer efficiency compared with other thermal recovery methods.

2. Classification of EOR Processes

The chemical recovery methods rely on using a wide variety of chemicals, such as surfactants, alkaline solutions, polymers, etc. The objectives of the chemical methods include reducing the oil/water interfacial tension, altering the rock wettability, and increasing the injected water viscosity to prevent viscous fingering and improve the overall oil sweep efficiency.

The miscible methods include the injection of a gas, such as carbon dioxide (CO_2), light hydrocarbon (CH_4), and nitrogen (N_2) into the reservoir. The injected gas completely mixes with the in-situ oil (at the first contact or after multiple contacts) when the reservoir pressure is greater than the minimum miscibility pressure (MMP). The objective of the miscible methods is to reduce the interfacial tensions between crude oil, water, and the rock, which significantly enhances the oil microscopic displacement efficiency. The CO_2 injection method for miscible oil recovery is gaining an increasing interest in the oil and gas industry because, in addition to its high oil recovery, some CO_2 could be retained or lost in the reservoir, thus decreasing its devastating impact as a greenhouse gas (GHG) on humans and the environment.

The other EOR methods include microbial (MEOR), nanoparticles, electromagnetic heating, and smart water injection. Despite their growing popularity, however, most of these emerging methods are still in the trial stage. Many of them are attracting research interest, but due to their implementation challenges in oil fields, they are not expected to impact the global oil production in the near future.

Nonetheless, it is well-known that oil production using EOR methods has been strongly correlated with economics and crude oil prices; however, the general trend of the EOR market has been increasing despite the tumultuous economic environments over the past decades. Thus, we believe that EOR market will continue to grow and play a key role in meeting the growing worldwide energy demand in the years to come despite the current environmental concerns and the ongoing transition to renewables, such as solar energy, biomass, and wind energy. We also believe that in the foreseeable future, renewables and nuclear energy alone will not be enough to replace the other primary energy sources and sustain or meet the growing energy demand of the steadily increasing world population.

DOI: http://dx.doi.org/10.5772/intechopen.105068

Author details

Hseen O. Baled* and Badie I. Morsi
University of Pittsburgh, United States of America

*Address all correspondence to: hob9@pitt.edu

References

[1] U.S. Energy Information Administration (EIA). International Energy Outlook 2021. 2021. Available from: https://www.eia.gov/outlooks/ieo/. [Accessed: January 26, 2022]

[2] United Nations. World Population Prospects 2019. 2019. Available from: https://population.un.org/wpp/. [Accessed: January 26, 2022]

[3] BP. Statistical Review of World Energy 2021. 2021. Available from: http://www.bp.com/statisticalreview. [Accessed: January 26, 2022]

[4] Organization of the Petroleum Exporting Countries (OPEC). World oil outlook 2021. 2021. Available from: https://www.opec.org/opec_web/en/publications/340.htm. [Accessed: January 26, 2022]

Section 2

Miscible Flooding

Chapter 2

Carbon Dioxide-Oil Minimum Miscibility Pressure Methods Overview

Eman Mohamed Ibrahim Mansour

Abstract

One of the essential parameters in carbon dioxide (CO_2) miscible flooding is the minimum miscibility pressure (MMP). Minimum miscibility pressure (MMP) is defined as the lowest pressure at which recovery of oil is (90–92%) at injection (1.2 PV) of carbon dioxide (CO_2). The injected gas and oil become a multi-contact miscible at a fixed temperature. Before any field trial, minimum miscibility pressure (MMP) must be determined. This parameter must be determined before any field trial because any engineer needs a suitable plan to develop an injection and surface facilities environment. Estimation of reliable (MMP) maybe by traditional laboratory techniques, but it is very costly and time-consuming. Also, it can rely on various literature **(MMP)** empirical correlations, but this is not a good strategy because each minimum miscibility pressure (MMP) correlation relates to a unique formation condition.

Keywords: enhanced oil recovery, CO_2 injection, minimum miscibility pressure (MMP), slim-tube test, computational method

1. Introduction

Miscible gas injection could enhance light oil reservoirs' enhanced recovery (EOR) process. Recycling CO_2 into oil formation reservoirs allows good gas storage in subsurface formations; consequently, the oil recovery will improve [1]. Miscible flooding project design mostly depends on his success in the minimum miscibility pressure correct determination [2]. The minimum flooding pressure reached the miscibility point, which is the maximum oil recovery achieved (90–92%) at the lowest pressure during injection (1.2 PV) of carbon dioxide (CO_2) [3]. Incremental oil recovery is negligible at a higher (CO_2) flooding project pressure than the MMP. In addition, recovery will sharply decrease at a pressure lower than MMP [4].

1.1 Importance of the minimum miscibility pressure (MMP)

- Minimum miscibility pressure (MMP) is an essential parameter in any design project of gas injection. At the minimum miscibility pressure (MMP), all oil was recovered within a porous medium of (CO_2).

- When oil and gas are miscible, displacement efficiency will be 100% [5].

1.2 Factors affecting minimum miscibility pressure (MMP)

oil composition, reservoir temperature, and (CO_2) purity effect on minimum miscibility pressure (MMP), where [6]:

- The minimum miscibility pressure (MMP) does not change as methane is present in the reservoir.
- As the oil gravity of oil becomes heavier as minimum miscibility pressure (MMP) increases. So, fields with heavy API are not suitable for carbon dioxide (CO_2)-EOR injection.
- As reservoir temperature is high, minimum miscibility pressure (MMP) is higher.
- Minimum miscibility pressure (MMP) is inversely related to the reservoir oil's C_5 to C_{30} summation.
- Minimum miscibility pressure (MMP) does not require methane to propane presence.

2. Methods of estimating (MMP)

There are several experimental, equation of state, and empirical equations for estimating (MMP).

2.1 Experimental methods for estimating minimum miscibility pressure (MMP)

Minimum miscibility pressure (MMP) estimates through several testing methods: slim-tube experiments, rising bubble experiments, multiple-contact experiments, and vanishing interfacial tension experiments [7].

2.1.1 Slim-tube experiment

These experiments are widely accepted experimental methods for estimating minimum miscibility pressure (MMP) because they can repeat the interaction of gas and oil in a one-dimensional porous medium [8]. As a result, that slim-tube experiment can replicate oil and gas interaction in a one-dimensional porous medium. It remains the most reliable method of estimating minimum miscibility pressure (MMP)[9]. The slim-tube basic test is the small-diameter tube packed with an unconsolidated porous medium [3]. It is an idealized medium for carbon dioxide (CO_2) and oil to develop dynamic miscibility [10]. The slim-tube experiment comes close to a one-dimensional displacement due to this large length-to-diameter ratio, thus the isolating of phase behavior affects the efficiency of removal [11]. A slim-tube is a long thin stainless steel-tube that is fifteen-meter, packed with sand or glass beads (commonly, sand packing is 160 to 200 mesh) [12]. The slim schematic-tube appears in **Figures 1** and 2.

The slim-tube is saturated with at least two PV light oils at the reservoir temperature. Then the system pressurized gradually to the operating pressure in the

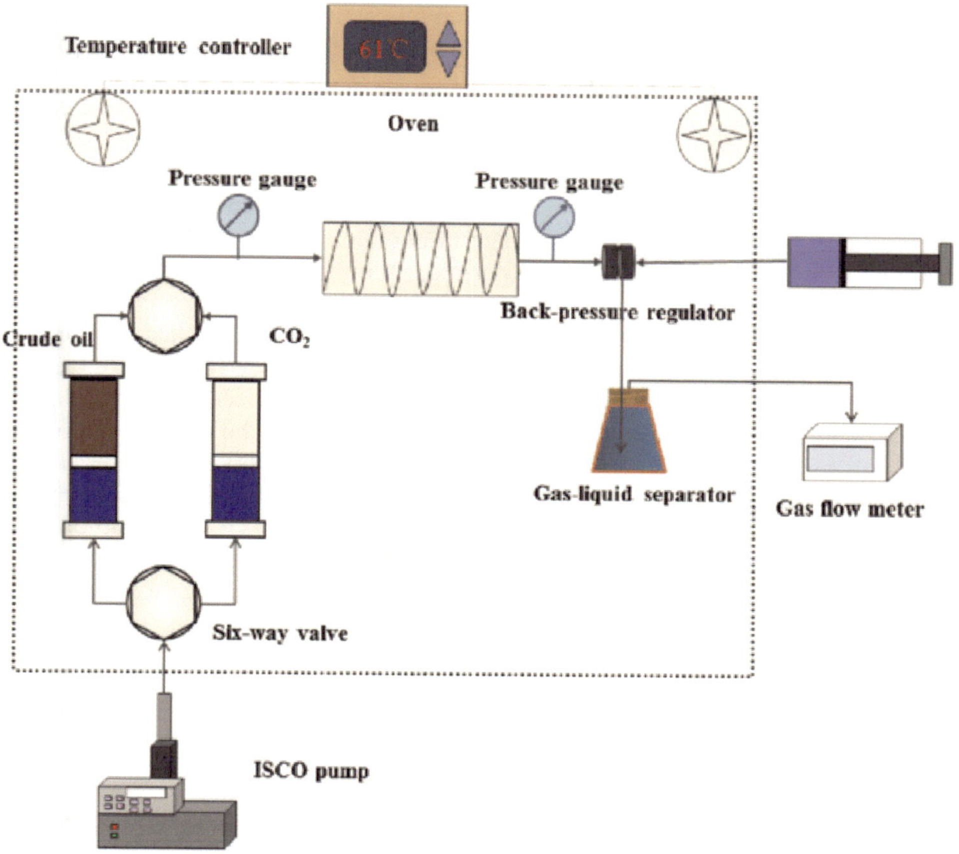

Figure 1.
Schematic diagram of the slim-tube test setup.

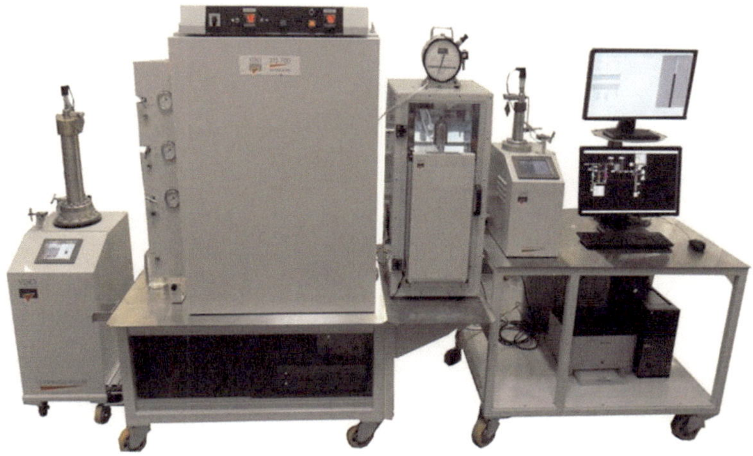

Figure 2.
Actual slim-tube test system.

backpressure regulator's presence [3]. It is pressure generally kept constant by a backpressure regulator. Upstream pressure changed with the backpressure regulator as it set [13]. To avoid pressure from one side of the diaphragm in the backpressure regulator from becoming significantly higher than the pressure on the other side and damaging the diaphragm [14]. It was required to pressurize the tube system gradually. Just the required pressure is reached, and the system will be equilibrated under this pressure. The carbon dioxide (CO_2) pressure pump was adjusted a little above the pressure of the backpressure regulator. The carbon dioxide (CO_2) flow rate is 4 to 8cm3/ hr. at a constant rate [15]. The experiment terminated when at least 1.2 PV of carbon dioxide (CO_2) was injected. Effluent flashed to atmospheric conditions, where the flow meter collects separator gas and the separated oil collects in a graduate cylinder. The pump's initial and final volumes of carbon dioxide (CO_2) were recorded to ensure the pump was not empty. The system was depressurized by venting loading gas gradually. After that, use two PV of methylene chloride in the slim-tube to remove residual oil and be ready for the next experiment [16]. The bubble point pressure of the formation oil must measure before the slim-tube experiment. On these results, the slim-tube test can repeat numerous times at different pressures greater than bubble-point pressure. The injected gas pour volume and oil recovery are recorded in each experiment. After indirect 1.2 gas pore volume, the minimum miscibility pressure (MMP) was observed from recovery data.

As shown in **Figure 3** [17].

2.1.2 Rising bubble experiment

Christiansen and Haines [18] are the first to introduce the rising bubble experiment as a fast option to the slim-tube test, where it can be measured within one hour. This method consists of eight inches high-pressure crystal clear-tube long packed with oil and set at a definite pressure and temperature. Gas introduces through the way of a needle at the tube bottom, forming a bubble and rising through the column [19]. This

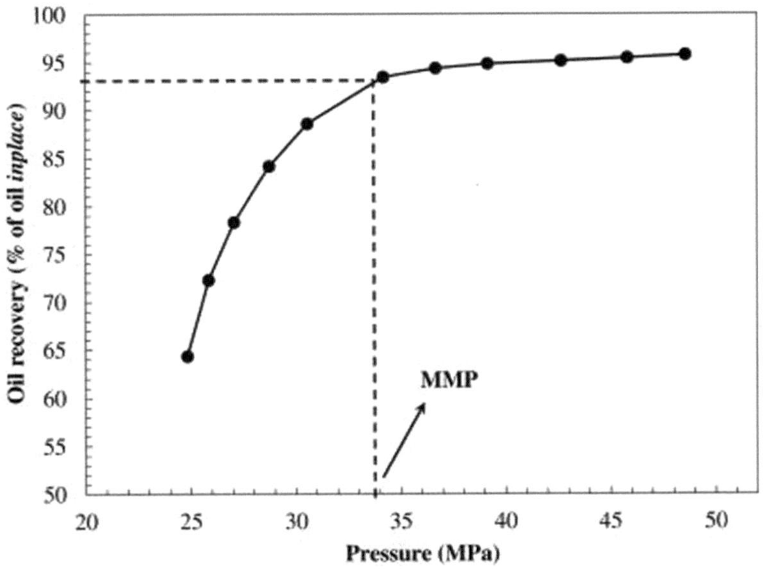

Figure 3.
Minimum miscibility pressure for (CO_2).

method can visually observe the miscibility between a gas bubble and an oil. The disadvantage of this method is significant limitations as not expensive and fast compared to the slim-tube method. This method is unreliable in predicting minimum miscibility pressure (MMP) in condensing drive and condensing/vaporizing gas drive (Mansour and Ragab, 2021) (**Figure 4**) [18].

2.1.3 Multiple-contact experiment

Multiple-contact experiments can detect minimum miscibility pressure (MMP) definite conditions. A multiple-contact test aims to study the gas and oil injection phase behavior [20]. The multiple-contact test is always on contacts between oil and gas. Oil and gas are mixed in a pressure-volume-temperature (PVT) cell [21]. A single PVT cell uses to make repeated contacts between oil and gas forward or backward. In a forward contract, the equilibrium gas retains after each contact.

In contrast, the equilibrium oil replaces with fresh oil—consequently, the equilibrium gas from the previous stage contacts fresh oil at each stage [22]. Equilibrium oil is retained in a backward connection, and the gas is replaced with new injection gas. The contacts are repeated till no change in the phase composition. These tests are repeated at different pressures until the repeated contacts result in a single phase (as shown visually from the cell window) [23]. The disadvantages of this method are that it can provide useful minimum miscibility pressure (MMP) and phase behavior data for gas floods that are purely condensing or vaporizing. Still, most gas flooding is condensing/evaporating drives, meaning that they have both condensing and drive features, but this makes the results of such experiments less accurate [24].

2.1.4 Vanishing interfacial tension experiment

Rao (1997) planned the vanishing interfacial tension (VIT) test as a technique for defining minimum miscibility pressure (MMP). This method involves a high temperature and pressure cell occupied with the CO_2 injection. A drop of crude oil (about 10% of the cell volume) is introduced through a capillary-tube [25]. It measures the interfacial tension (IFT) between oil and CO_2 injected gas at numerous pressures and a specific temperature. The analyzing data were determined by the shape of the hanging oil drop and the oil and gas densities. The pressure increases by pumping more gas into the cell, and the interfacial tension (IFT) measurement is repeated

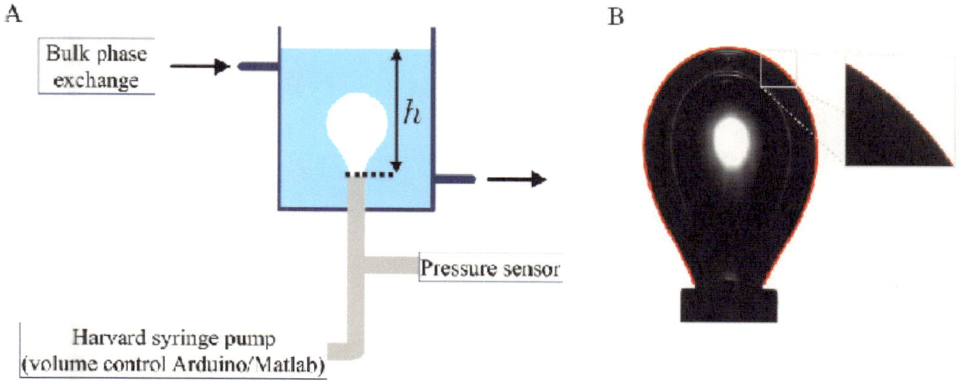

Figure 4.
Rising bubble experiment.

several times [26]. The minimum miscibility pressure (MMP) is approached by extrapolating the interfacial tension (IFT) plot versus pressures. The disadvantage of this test is that after extending their analysis to a multi-component mixture, and concluding that vanishing interfacial tension (VIT) experiments may not be a dependable method of determining minimum miscibility pressure (MMP) [27]. Among these, all experimental methods, the only known experiment of minimum miscibility pressure (MMP) between oil and injection gas is the slim-tube testing.

The experimental methods for estimating minimum miscibility pressure (MMP) have much money and time-overwhelming disadvantages. But these tests can provide us with useful phase behavior data that can be used to estimate and confirm the reliability of a computational minimum miscibility pressure (MMP) [26].

2.2 Computational method of estimating minimum miscibility pressure (MMP)

Empirical correlations for approximating minimum miscibility pressure (MMP) provide fast and cheap alternatives to experimental methods. It is beneficial for quick screening reservoirs for potential carbon dioxide (CO_2) flooding. Various empirical correlations for estimating minimum miscibility pressure (MMP) have been calculated from regression data analysis of slimtube data [28]. Generally, empirical correlations for the predicting of minimum miscibility pressure (MMP) reservoir temperature, the (C2-C6) content of reservoir fluid, and API (oil gravity) as input parameters [29]. This study includes popular minimum miscibility pressure (MMP) empirical correlations reported in the petroleum literature are included in this study. It can be used as a practical guide for the application of different reservoir oils, such as Cronquist [30], Lee [31], Yelling and Metcalfe [32], Alston et al. [33], Emera and Sarma [34], Liao et al. [35], and Mansour et al. [36].

2.2.1 Cronquist empirical correlation

Cronquist [30] empirical correlation is based on the reservoir temperature, pentane plus (C_5^+) molecular weight, and volatile oil fraction as (CH4 and N2) for minimum miscibility pressure (MMP) estimation as shown in Eq. (1) [30].

$$MMP = 0.11027 + (1.8T_R + 32)^y \tag{1}$$

where $Y = 0.744206 + 0.0011038 \times MWTC_{5+} + 0.0015279 \times Vol.$

The experimental data range tested used in this study is as the following:

- The oil gravity (API) ranged from 23.7 to 44 8.
- The temperature ranged from 21.67 to 120.8°C.
- The experimental (MMP) ranged from 7.4 to34.5 MPa.

2.2.2 Lee empirical correlation

Lee [31] predicted a model to estimate minimum miscibility pressure (MMP) using reservoir temperature as input data only by considering carbon dioxide (CO_2) vapor pressure, as shown in Eq.(2). If any reservoir oil's bubble point pressure (BP) is more than minimum miscibility pressure (MMP), the bubble point pressure (BP) takes as minimum miscibility pressure [31]. The bubble point can be detected from the constant mass study test [37].

$$MMP = 7.3924 \times 10^{b} \quad (2)$$

where b = 2.772−(1519/(492 + $1.8T_R$)).

2.2.3 Yelling and metcalfe empirical correlation

Yelling and Metcafe (1980) proposed an empirical correlation for estimating minimum miscibility pressure (MMP) at different reservoir temperatures by using the equation Eq. (3). This correlation is not dependent on oil composition and is based only on reservoir conditions. The empirical correlation of minimum miscibility pressure (MMP) is varied from 15 to 19 Mpa approximately [32].

$$\begin{aligned} MMP = {} & 12.6472 + 0.015531 \times (1.8T_R + 32) + 1.24192 \times 10^{-4} \\ & \times (1.8T_R + 32)^2 - 716.9427/(1.8T_R + 32) \end{aligned} \quad (3)$$

The limitation of reservoir temperature data used 35.8 °C < TR < 88.9°C. Suppose the minimum miscibility pressure (MMP) is less than the bubble point pressure (BP) for any sample. The bubble point pressure is taken as the minimum miscibility pressure (MMP) determined by the constant mass study test [38].

2.2.4 Alstonetal et al. empirical correlation

Alston et al. [33] presented an empirical correlation for minimum miscibility pressure (MMP) caused by gas solution in reservoir fluids. The minimum miscibility pressure empirical correlation that is in Eq.(4) was predicted based on carbon dioxide (CO_2) composition stream, light oil fraction (CH4 + N2), reservoir temperature, pentane plus (C_5^+) molecular weight, and Intermediate oil fraction (C_2 to C_4, H_2S, and CO_2). So, they proposed an impurity factor for predicting minimum miscibility pressure empirical correlation (MMP) by contaminated or en-riched carbon dioxide (CO_2) stream (Alston, Kokolis, and James, 1985).

$$p_{CO2} = 1.25 * 10^{-7}\,(1.8t - 460)^{1.06}(M_{WC5+})^{1.78}\left(\frac{x_{vol}}{x_{int}}\right)^{0.136} \quad (4)$$

Also, suppose the minimum miscibility pressure (MMP) of volatile reservoir oil is fewer than the saturation pressure (BP). In that case, the saturation pressure act as the minimum miscibility pressure (MMP).

2.2.5 Emera and Sarma empirical correlation

Emera and Sarma (2005) presented genetic logarithm (GA)-depending on correlation to predicate minimum miscibility pressure (MMP) as shown in Eq.(5). The input data parameters that are based on this correlation are (C_1 and N_2) volatiles ratio, reservoir temperature, intermediates components (C_2–C_4, H_2S, and CO_2), pentane plus (C_5^+) molecular weight, and (C_2–C_4, H_2S, and CO_2) [34].

$$MMP_{pure} = 0.003 * T^{0.544} \left(MW_{C_{5+}}\right)^{1.006} \left(Y_{VOL}|Y_{INT}\right)^{0.143} \quad (5)$$

2.2.6 Liao et al. empirical correlation

Liao et al. [35] offered minimum miscibility pressure (MMP) empirical correlation depending on (CH4 + N2) oil fraction, (C_2 to C_4, H_2S, and CO_2) intermediate oil fraction, pentane plus (C_5^+) molecular weight, and reservoir temperature, as shown in Eq.(6).

$$MMP_{pure} = 0.003 * T^{0.544} \left(MW_{C_{5+}}\right)^{1.006} \left(Y_{VOL}|Y_{INT}\right)^{0.143} \quad (6)$$

This minimum miscibility pressure (MMP) empirical correlation was appropriate for low permeability reservoirs. The characteristics of oil reservoir low permeability must be (vol/yint>1), where the experimental and published data were used in this correlation [35].

2.2.7 Mansour et al. empirical correlation

Mansour et al. [36] proposed a new method for predicting (MMP) of a multi-component volatile oil reservoir. This model used twenty-live crude oil samples to correlate this unique formula with new constants. The data range for using this equation API from 40.5 to 26 and in range temperature from 160 to 246°F. The developed (MMP) equation gives good results to reduce the previous correlations errors, where (E_r) was found to be 0.627%, (E_a) 2.4%, (E_r) 0.627%, (R_2) 94.82% (S), and 2.7%. Consequently, this new model has better accuracy than previous literature correlations, as shown in Eq. (7) [36].

$$\mathrm{Ln(MMP)} = \beta_0 + \beta_1(\mathrm{LnT}) + \beta_2(\mathrm{LnMWT}_{C5}{}^{+}) + \beta_3(\mathrm{yVol/yint}) \quad (7)$$

where β_0, β_1, β_2, β_3 are the coefficient values and have the following values 11.222, −0.355, −0.2069, and 0.039, respectively.

Nomenclature

API	American Petroleum Institute, degree
BP	Bubble point(saturation pressure), psi
CO_2	Carbon Dioxide
CMD	Constant Mass Depletion
E_a	Average Absolute percent relative error
E_r	Average percent relative error %
PV	Pore Volume
MMP	Minimum Miscibility Pressure
MW_{C5}^{+}	Molecular weight of pentane plus
R^2	Correlation Coefficient
S	Standard deviation
T	Reservoir temperature
y	Mole percentage
y_i	Mole fraction of I component

Subscript

Int	Intermediate
Psi	Pound/square inch
Res	Reservoir
vol	Volatile

Author details

Eman Mohamed Ibrahim Mansour[1,2]

1 PVT Lab, Production Department, Egyptian Petroleum Research Institute, Cairo, Egypt

2 PVT Service Center, Egyptian Petroleum Research Institute, Cairo, Egypt

*Address all correspondence to: emanmansour84@yahoo.com

References

[1] Rezaei M, Eftekhari M, Schaffie M, Ranjbar M. A CO2-oil minimum miscibility pressure model based on multi-gene genetic programming. Energy Exploration & Exploitation. 2013;**31**:607-622

[2] Ragab A, Mansour EM. Investigating the impact of PVT analysis errors on Material balance calculations for oil reservoirs. Petroleum & Coal. 2020

[3] Mansour EM, Al-Sabagh AM, Desouky SM, Zawawy FM, Ramzi MR. Experimental approach of minimum miscibility pressure for CO2 miscible flooding: Application to Egyptian oil fields. International Journal of New Technology and Research. 2016;**2**:263507

[4] Li L, Zhang Y, Sheng JJ. Effect of the injection pressure on enhancing oil recovery in shale cores during the CO2 huff-n-puff process when it is above and below the minimum miscibility pressure. Energy & Fuels. 2017;**31**:3856-3867

[5] Perera MS, Anne RP, Gamage TD, Rathnaweera AS, Ranathunga AK, Choi X. A review of CO2-enhanced oil recovery with a simulated sensitivity analysis. Energies. 2016;**9**:481

[6] Abdullah N, Hasan N. Effects of miscible CO2 injection on production recovery. Journal of Petroleum Exploration and Production Technology. 2021;**11**:3543-3557

[7] Golkari A, Riazi M. Experimental investigation of miscibility conditions of dead and live asphaltenic crude oil–CO_2 systems. Journal of Petroleum Exploration and Production Technology. 2017;**7**:597-609

[8] Ahmad W, Vakili-Nezhaad G, Al-Bemani AS, Al-Wahaibi Y. Experimental determination of minimum miscibility pressure. Procedia Engineering. 2016;**148**:1191-1198

[9] Saini D, Rao DN. Experimental determination of minimum miscibility pressure (MMP) by gas/oil IFT measurements for a gas injection EOR project. In: SPE Western Regional Meeting. 2010

[10] Mansour EM, Al-Sabagh AM, Desouky SM, Zawawy FM, Ramzi M. A laboratory investigation of carbon dioxide-enhanced oil recovery by focusing on CO2-oil physical properties. Egyptian Journal of Petroleum. 2019;**28**: 21-26

[11] Abdurrahman M, Bae W, Permadi AK. Determination and evaluation of minimum miscibility pressure using various methods: Experimental, visual observation, and simulation. Oil & Gas Science and Technology–Revue d'IFP Energies nouvelles. 2019;**74**:55

[12] Bui LH. Near Miscible CO2 Application to Improve Oil Recovery. University of Kansas; 2010

[13] Ragab A, Mansour EM. Enhanced Oil Recovery: Chemical Flooding. London, UK: IntechOpen; 2021

[14] Cao M, Yongan G. Oil recovery mechanisms and asphaltene precipitation phenomenon in immiscible and miscible CO2 flooding processes. Fuel. 2013;**109**:157-166

[15] Moradi S, Dabiri M, Dabir B, Rashtchian D, Emadi MA. Investigation of asphaltene precipitation in miscible gas injection processes: Experimental study and modeling. Brazilian Journal of Chemical Engineering. 2012;**29**:665-676

[16] Bahadori A. Fundamentals of Enhanced Oil and Gas Recovery from Conventional and Unconventional Reservoirs. Gulf Professional Publishing; 2018

[17] Oyinloye O, Al Darmaki N, Al Zarooni M, Boukadi F, Nantongo H. Estimation of minimum miscibility pressure for flue gas injection using soft experimentations. Natural Resources. 2021;**12**:363-380

[18] Christiansen RL, Haines HK. Rapid measurement of minimum miscibility pressure with the rising-bubble apparatus. SPE Reservoir Engineering. 1987;**2**:523-527

[19] Zhou D, Orr FM. An analysis of rising bubble experiments to determine minimum miscibility pressures. In: SPE Annual Technical Conference and Exhibition. 1995

[20] Yang FL, Yu P, Xue Z. Multiple-mixing-cell model for calculation of minimum miscibility pressure controlled by tie-line length. In: Geofluids. 2020

[21] Amao AM, Siddiqui S, Menouar H, Herd BL. A new look at the minimum miscibility pressure (MMP) determination from slimtube measurements. In: SPE Improved Oil Recovery Symposium. 2012

[22] Sun H, Li H. A modified cell-to-cell simulation model to determine the minimum miscibility pressure in tight/shale formations. Oil & Gas Science and Technology–Revue d'IFP Energies Nouvelles. 2021;**76**:48

[23] Zhao G-B, Adidharma H, Towler B, Radosz M. Using a multiple-mixing-cell model to study minimum miscibility pressure controlled by thermodynamic equilibrium tie lines. Industrial & Engineering Chemistry Research. 2006;**45**:7913-7923

[24] Zhang K, Jia N, Zeng F, Luo P. A new diminishing interface method for determining the minimum miscibility pressures of light oil–CO2 systems in bulk phase and nanopores. Energy & Fuels. 2017;**31**:12021-12034

[25] Orr J, Franklin M, Jessen K. An analysis of the vanishing interfacial tension technique for determination of minimum miscibility pressure. Fluid Phase Equilibria. 2007;**255**:99-109

[26] Teklu TW, Alharthy N, Kazemi H, Yin X, Graves RM. Vanishing interfacial tension algorithm for MMP determination in unconventional reservoirs. In: SPE Western North American and Rocky Mountain Joint Meeting. 2014

[27] Seyyedsar SM, Assareh M, Ghotbi C. Minimum miscibility pressure calculation: A comparison between multi-contact test and compositional simulation. In: CIVILICA, 14th Iranian National Chemical Engineering congress. Tehran; 2012

[28] Delforouz FB, Movaghar MRK, Shariaty S. New empirical correlations for predicting minimum miscibility pressure (MMP) during CO_2 injection; implementing the group method of data handling (GMDH) algorithm and Pitzer's acentric factor. Oil & Gas Science and Technology–Revue d'IFP Energies Nouvelles. 2019;**74**:64

[29] Alomair O, Malallah A, Elsharkawy A, Iqbal M. Predicting CO2 minimum miscibility pressure (MMP) using alternating conditional expectation (ACE) algorithm. Oil & Gas Science and Technology–Revue d'IFP Energies nouvelles. 2015;**70**:967-982

[30] Cronquist C. Carbon dioxide dynamics displacement with light reservoir oil. In: DOE Annual Symposium. Tulsa; 1978

[31] Lee IJ. Effectiveness of Carbon Dioxide Displacement under Miscible and Immiscible Conditions. 1979

[32] Yellig WF, Metcalfe RS. Determination and prediction of CO2 minimum miscibility pressures (includes associated paper 8876). Journal of Petroleum Technology. 1980;**32**: 160-168

[33] Alston RB, Kokolis GP, James CF. CO2 minimum miscibility pressure: A correlation for impure CO2 streams and live oil systems. Society of Petroleum Engineers Journal. 1985;**25**:268-274

[34] Emera MK, Sarma HK. Genetic Algorithm (GA)-based correlations offer more reliable prediction of CO_2-oil physical properties. In: Canadian International Petroleum Conference. Calgary, Alberta, Canada; 2005

[35] Liao C, Liao X, Chen J, Ye H, Chen X, Wang H. Correlations of minimum miscibility pressure for pure and impure CO2 in low permeability oil reservoir. Journal of the Energy Institute. 2014;**87**: 208-214

[36] Mansour EM, Al-Sabagh AM, Desouky SM, Zawawy FM, Ramzi M. A new estimating method of minimum miscibility pressure as a key parameter in designing CO2 gas injection process. Egyptian Journal of Petroleum. 2018;**27**: 801-810

[37] Mansour EM, Desouky SM, Batanoni MH, Mahmoud MR, Farag AB, El-Dars FS. Modification proposed for SRK equation of state. Oil & Gas Journal. 2012;**110**:78-78

[38] Mansour EM, Farag AB, El-Dars FS, Desouky SM, Batanoni MH, Mahmoud MRM. Predicting PVT properties of Egyptian crude oils by a modified soave–Redlich–Kowng equation of state. Egyptian Journal of Petroleum. 2013;**22**: 137-148

Chapter 3

Miscible Displacement Oil Recovery

Nasser Mohammed Al Hinai and Ali Saeedi

Abstract

Miscible gas injection (MGI) is an effective enhanced oil recovery (EOR) method used worldwide often for light oil recovery. In the petroleum industry, many MGI processes typically involve injection of an associated gas (AG) mixture or CO_2, which have both been recognised as excellent candidates for such processes. The initial part of this chapter provides a broad introduction and background to the EOR techniques used worldwide as well as those implemented in Oman oil fields and briefly discusses their critical importance. Oman is one of the most active countries in terms of successful MGI processes in the Middle East, hence the emphasis given in this chapter to such projects in this country. The second part covers the technical details of the MGI process and the potential problems and challenges associated with it, while the third part focuses mainly on the common techniques used to control gas mobility during gas flooding including MGI. The impediments and challenges for wider application of the mobility control techniques are also covered. The last section presents a sample feasibility evaluation for a real oil field around the implementation of mobility control techniques for an MGI process.

Keywords: miscible gas injection, enhanced oil recovery, gas mobility control

1. Introduction

Over the past few decades, the rate of the new substantial oil discoveries has been on the decline. As a result nowadays, many oil companies are trying to maximise oil production from their existing reserves and maintain oil flow rates at or above the economic level through production optimisation and the use of enhanced oil recovery (EOR) techniques [1]. Enhanced oil recovery refers to the methods of increasing or maintaining the ability of oil to flow through interconnected pores towards the production wells by changing the physical and/or chemical properties of the in-situ fluid-rock system. Presently, the average recovery factor (RF) from mature oilfields under the primary and secondary recovery is only 20–40% [2]. Given the earlier mentioned lack of substantial new discoveries, increasing the RF from matures fields has become important to meet the growing energy demand in the years to come.

During the life cycle of an oil field, the oil extraction may occur typically in three recovery stages of primary, secondary and tertiary (i.e. EOR). Essentially, the petroleum product is produced from the reservoir initially by the natural reservoir energy such as the solution gas drive, gas cap drive and aquifer influx [3]. This is often termed as primary recovery, where the first wells drilled in the field are able to produce the oil from the reservoir without any intervention. In this stage primarily, the pressure gradient between the reservoir and surface controls the hydrocarbon flow into the well and then to surface. Over time, the reservoir pressure may decline reducing the pressure at the bottom hole which may then become closer to the hydrostatic head of the fluid column in a production well reducing the oil flow rate achievable from the well. Subsequently, secondary recovery methods may be applied, for example, by injecting water or gas via injection wells into the reservoir to maintain the reservoir pressure and eliminate or minimise the previously observed decline in oil flow. This type of recovery methods has its own technical and economic limitations as may be determined by the cost and availability of injection fluids and/or the issues that may arise during the development of the in-situ flooding. For instance, in both water and gas flooding, the difference in fluid properties between the displacing fluids and to be displaced in-situ oil can result in unstable displacement, leading to a large oil volume left behind due to poor displacement efficiency and early breakthrough. Therefore, the application of such techniques may typically add up to only 40–50% of eventual oil recovery.

When the oil in a reservoir can no longer be produced by natural reservoir pressure (i.e. primary recovery), or by water or immiscible gas injection (i.e. secondary, improved recovery methods (IOR) or pressure maintenance), EOR techniques may be considered. In general, as briefly referred to earlier, EOR techniques aim to stimulate oil flow by overcoming the physical, chemical and geologic factors that inhibit the production of the remaining hydrocarbons [4]. One of the most widely implemented EOR processes today is thermal recovery, which involves heating the oil bearing interval with steam or hot water to reduce the oil viscosity. Miscible gas injection (MGI) is another most widely used approach today, which is carried out by the injection of a high-pressure gas, such as carbon dioxide or hydrocarbon-associated gas, to sweep additional oil towards the wellbores by employing a number of in-situ mechanisms such as oil viscosity and IFT reductions. Chemical agents dissolved in water and injected into the reservoir can also improve the displacement properties during a water flood. Currently, various EOR projects executed around the world, as shown in **Figure 1**, account for only 3.5% (3 million barrels per day (MMbpd)) of the

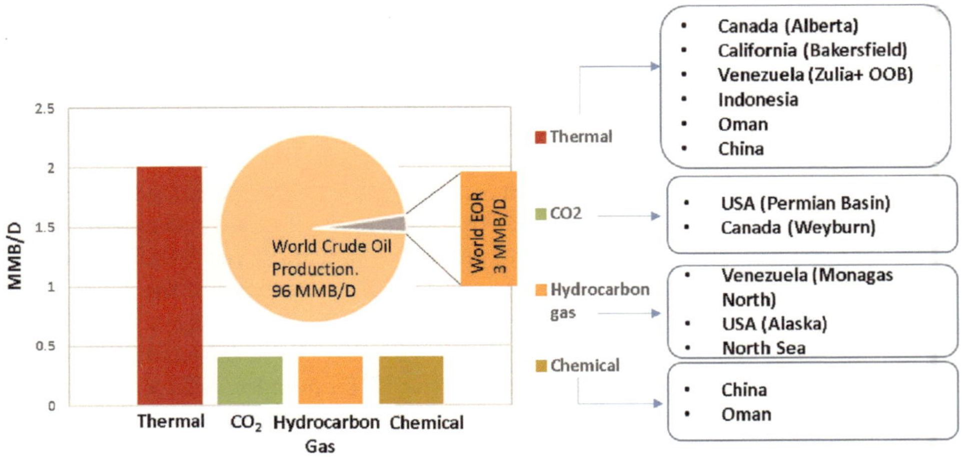

Figure 1.
Worldwide EOR projects contribute to global oil production [5].

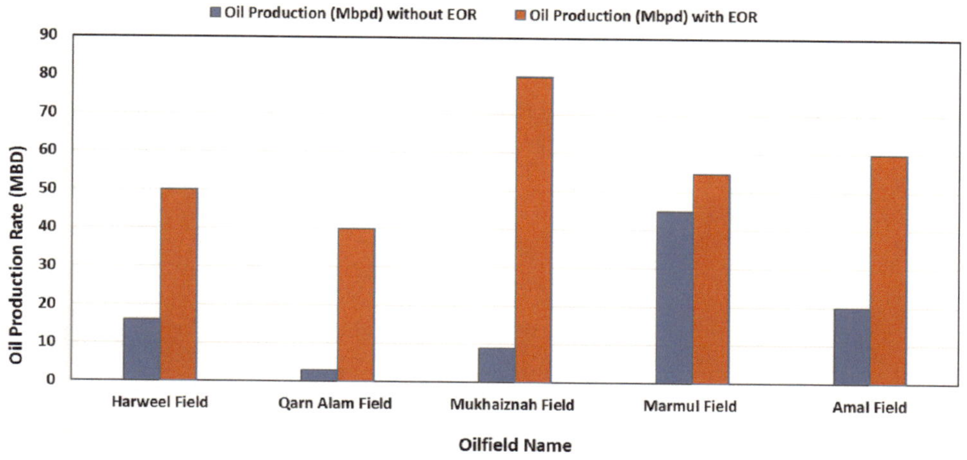

Figure 2.
The contribution of current EOR projects implemented in Oman oil fields [7].

total world oil production (96 MMbpd) [5]. However, further application of these technologies has the potential to increase oil recovery from existing fields and new discoveries and alleviate oil supply shortage in the future [5, 6].

The application of EOR techniques in Oman may be considered as a successful example of how such techniques may be used to boost oil production and achieve substantial enhancements in recovery. Over the past decade, a number of EOR projects in the Middle East (ME) have been executed. Among the ME countries, Oman leads the way mainly owing to its declining overall oil production rate [7] which has seen EOR to become a major strategy to meet target oil production from its existing fields [8]. In 2007, this country's oil production declined to an average of 700,000 bpd. However, with the aid of EOR methods, the field operators have been able to increase the country's overall oil production to its current level of nearly 1 million bpd. Miscible gas injection is one of the EOR techniques used in the country. The largest fields produced using EOR techniques in Oman and the indicative contributions made by such techniques in each field are depicted in **Figure 2**. The daily oil production rate from these fields with implanting EOR techniques varies between 40 and 80 thousand bpd (Mbpd). While without EOR, it was 3–45 Mbpd making such techniques the key driver of Oman's oil production nowadays [9].

2. Miscible gas injection (MGI)

The MGI is one of the most effective EOR methods used to enhance the production of light crude oil in the petroleum industry [10]. This method is PVT driven in which the injected gas (CO_2, associated gas (AG) or natural gas liquids (NGL)), in addition to helping with pressure maintenance, would mix with and alter the properties of the in-situ oil allowing the otherwise trapped oil to become mobile and easily displaced [11, 12]. During the miscible gas flooding, the injected gas would become miscible with the reservoir oil at or above the minimum miscibility pressure (MMP) of the reservoir oil (**Figure 3**). By definition, the MMP is the pressure at which the mass transfer and molecular interactions between the gas and oil intensify forcing the physical and chemical properties of the two phases to converge [14, 15]. In other

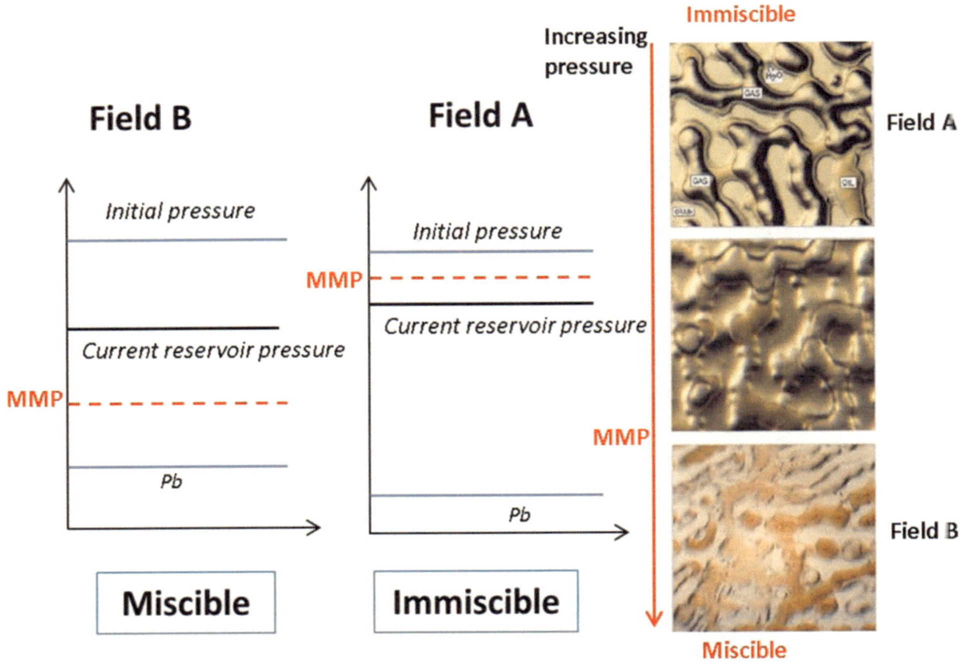

Figure 3.
Development of miscibility of injected CO_2 in oil at miscible and immiscible pressures [13].

words, upon reaching MMP, the gas acts as a solvent for the oil towards forming a single fluid phase (liquid) in the reservoir with the potential of effectively reducing the saturation of the remaining oil to near zero under ideal conditions [4]. During this process, the improved displacement efficiency of the flood is realised via three main mechanisms including substantial reduction in IFT (i.e. elimination of the interface between the two fluids and reduction of capillary pressure to zero), reduction of oil viscosity and oil swelling [16, 17]. The value of MMP depends on the reservoir temperature as well as the compositions of the injected gas and in-situ oil [18, 19].

In general, the miscibility process of the crude oil-gas system may occur through two paths of multi-contact miscibility (MCM) and first-contact miscibility (FCM) [20]. The MCM would take place if the in-situ pressure is equal to MMP which, as discussed previously, is a critical property to be taken into account for designing an MGI process [21]. The MCM may develop gradually via a number of processes including vaporising gas drive, condensing gas drive and a combination of the two drives [15]. On the other hand, when the reservoir pressure is adequately high and well above MMP, FCM would take place in which the injected gas would develop miscibility with the in-situ oil at all proportions as soon as they are brought in contact. Since FCM would only occur at high enough pressures, depending on the type of injectant used, achieving this type of miscibility could be challenging.

The type of the injected gas used for MGI depends on the gas availability and reservoir conditions [2] with the common gases used around the world being CO_2, hydrocarbon gas mixture (AG, NGL), flue gas and N2 [20, 22]. Carbon dioxide, which has been most widely used in the United State, Canada and China [23], can achieve miscibility at relatively low pressures (when compared with other gases) and has a relatively high density (can be similar to oil). The latter can help to reduce the severity of gravity segregation and override which can negatively affect the sweep efficiency.

The use of this gas for flooding can also help to reduce the global level of CO_2 emissions. However, some of the main challenges for a successful CO_2 flooding in general are the availability of CO_2 and corrosion in wells and surface facilities, which can result in considerable cost increases, in particular, for remotely located fields.

In the Middle East, the available CO_2 supply is limited to those associated with large industrial sources [24] which, when combined with the earlier mentioned issues associated with using this gas, has made its wide application limited. However, the hydrocarbon gas injection could be considered for MGI processes more widely for which the produced AG is usually readily available from the field itself or those close by. On the other hand, as mentioned earlier, unlike CO_2, conducting MGI using AG, depending on the gas composition, requires a relatively high pressure to achieve miscibility. To date, there have been three MGI projects (at either pilot- or field-scale) in the Middle East as reported in the literature [25]. Two of the projects involve miscible CO_2 injection and the other utilises AG injection. The Rumaitha Field in Abu Dhabi was the first pilot miscible CO_2 injection implemented in the region [25–28]. The second pilot CO_2-EOR project has been implemented in Minagish Oolite Reservoir in west Kuwait [29]. The third project has been implemented in Field A located in the Harwell Cluster in southern Oman in which the field's AG mixture (CH_4 enriched with light and heavy hydrocarbon fractions found in natural gas as well as considerable amounts of sour gases (3–5 mol % H_2S and 10–25 mol % CO_2)) is used for reinjection [30, 31]. As will be discussed in further details later with field case for improving the MGI process in this field.

Harweel Fields consist of a cluster of reservoirs deep within the tight carbonate oil-bearing rocks in the south of Oman in the Petroleum Development Oman (PDO) concession area, as shown in **Figure 4**. The figure also presents a geological cross section of the carbonate stringers, as encased in the Ara salt and the general geological setting of the area. The fields are expected to make a significant contribution to the Sultanate's oil production over the coming 30 years. The reservoir rocks in these fields are more than half a billion years old (where the hydrocarbon deposits are among the oldest in the world) located at a depth of about 5 km, making them PDO's deepest producing oil fields [33, 34]. As indicated earlier, the MGI in Field A, located in the cluster, has already begun in which the source of the injection gas is the Field's AG [30]. The produced AG mixture is reinjected into the reservoir at high pressures of up

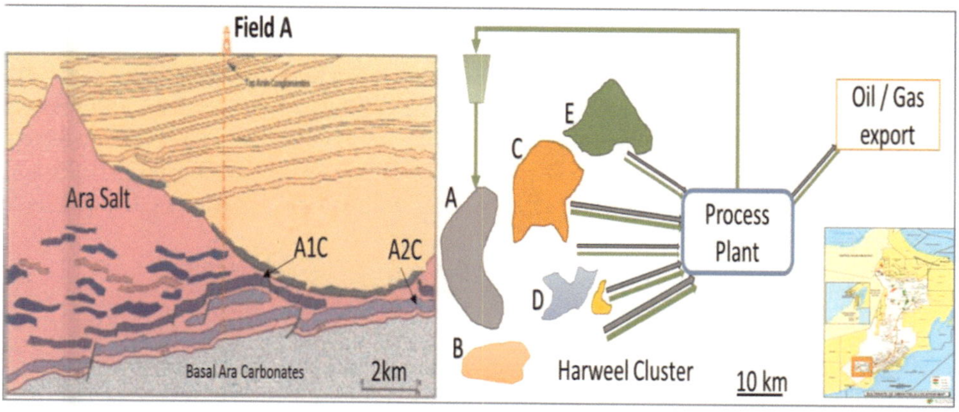

Figure 4.
Geological cross section of the carbonate stringers (left) and an aerial overview of Harweel Fields in southern Oman (right) [32].

to 45 MPa during which the injected gas develops miscibility with the in-situ oil under the reservoir's high temperature (up to 377 K). The reservoir contains a light crude oil with a typical gravity of 42°API and a viscosity of 0.23 cP at reservoir conditions. It was initially estimated that up to 47% of the Field's original oil in place (OOIP) could be recovered with the MGI process [33]. However, it has been realised since then that the presumed RF might not be eventually achievable due to the technical and operational challenges faced in this field, e.g. premature gas breakthrough and high degree of reservoir heterogeneity. As mentioned earlier, this chapter will be mainly focusing on addressing some of the technical challenges experienced during MGI in Field A and similar fields by proposing and testing a novel mobility control technique applicable to such a high-pressure and temperature environment.

3. Challenges associated with MGI process

As with other EOR techniques, MGI can be economically expensive and technically challenging to implement [35]. For example, even when the injection gas is readily available on site (e.g. associated gas), gas processing, handling and compression as part of the expected gas recycling scheme can be costly [2, 36]. Full life cycle economics of a gas injection project, therefore, must be taken into account to justify its implementation. In addition, as an example, a technical challenge in achieving a profitable MGI is the instability of the oil displacement process in the reservoir mainly due to the expected unfavourable mobility ratio and possible gravity segregation whose effects may be intensified by the level of reservoir heterogeneity.

From a more general technical perspective, the efficiency of an MGI is controlled by the collective effects of several physical forces acting on the displacement front. These forces include the viscous forces that stem from viscosity contrast in the flood, gravity forces caused by fluid-fluid density differences, dispersive forces driven by the fluid concentration gradients and, finally, the capillary forces that have roots in the IFT between any immiscible fluids. The large differences in fluid viscosities can cause viscous fingers at the displacement front If the vertical permeability in the reservoir is quite high, a pronounced density difference can cause gravity segregation. Both of the above have the potential to leave a large amount of oil unswept. The capillary and dispersive forces tend to enhance the fluid mixing but do not often overwhelm the viscous fingering [37, 38]. Therefore, the gravity and viscosity forces are the essential forces driving the instability of the oil displacement process during MGI [39]. Provided in the following two subsections are further details about the underlying mechanisms behind these two forces and how they may interfere with the performance of an MGI process. Possible mitigation strategies to lessen their effects will be outlined and adequately discussed in later sections of this chapter.

3.1 Viscous fingering

When a fluid is injected into a reservoir to displace another, there is almost never a collective perfect piston-like displacement across the entire reservoir interval. Especially in a gas flood, unstable displacement due to viscous fingering can lead to uneven or poor sweep, as depicted in **Figure 5** [40]. Viscous fingering is generally defined as a hydrodynamic instability that occurs between two fluids of differing mobility/

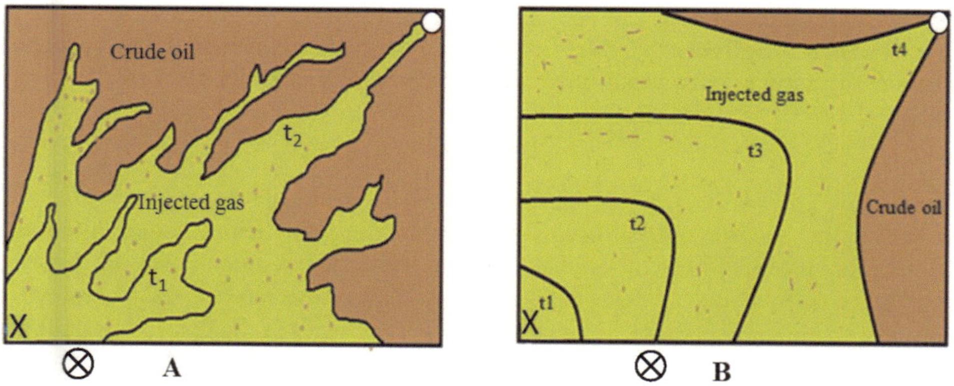

Figure 5.
Effect of viscous fingering on the development of areal sweep efficiency against time (t) in a quarter of a five-spot flood pattern during gas flooding, (A) an unstable displacement with poor macroscopic sweep, (B) a stable displacement good with macroscopic sweep, (⊗) an injection well and (O) a production well [40].

viscosity in the porous media that could lead to reduced sweep efficiency and early breakthrough [39, 41, 42]. The terms mobility, mobility ratio and that in a gas flood mobility ratio may be interchangeably used with viscosity ratio would be defined and discussed shortly. In MGI, there are several parameters that affect the viscous instability at the fluid-fluid interface including fluid viscosities, degree of miscibility, gas dissolution and exsolution and reservoir heterogeneity [39, 42–44]. However, the viscosity contrast and permeability heterogeneity are the two that mainly control the dynamics of the fingering phenomenon [37, 45]. The importance of mobility/viscosity ratio may be further realised after defining the mobility ratio (M) as a widely used criterion to characterise and determine the occurrence and possible effects of viscous fingering.

As indicated by Eq. (1), the mobility of a fluid (λ_i) in a porous medium may be defined as the ratio of effective permeability (K_i) and effective viscosity (μ_i) experienced by the fluid while flowing in the medium [40, 46, 47],

$$\lambda_i = \frac{K_i}{\mu_i} \tag{1}$$

Subsequently, for any fluid-fluid displacement, such as an MGI, the mobility ratio (M) can be simply defined as the mobility of the displacing fluid over that of the displaced fluid [40, 47]. For instance, Eq. (2) defines M for an MGI process where gas displaces the in-situ oil.

$$M = \frac{\lambda_{gas}}{\lambda_{oil}} = \frac{K_{gas}/\mu_{gas}}{K_{oil}/\mu_{oil}} \tag{2}$$

Where μ_{oil} and μ_{gas} are the oil and injected gas viscosities, respectively. For a miscible displacement, where the gas solvent may displace the oil at irreducible water saturation and the effective permeability to both fluids may be considered to be the similar, Eq. (2) may be reduced to Eq. (3) [40]. Furthermore, during gas flooding, due to the large viscosity contrast between the gas and the in-situ oil, viscosity ratio may be considered adequate for qualitative evaluation of viscous instability in the flood [40, 45, 48]. Therefore, for the purpose of qualitatively characterising the effect of viscous fingering on the performance of an MGI process, the viscosity ratio may be used interchangeably with the mobility ratio [40].

$$M = \frac{\mu_{oil}}{\mu_{gas}} \tag{3}$$

During its development, the severity of viscous fingering increases with increase in the mobility/viscosity ratio of the fluid system. If M is larger than unity, the displacement becomes unstable resulting in the development of viscous fingers. Therefore, to achieve a stable displacement, where possible, the viscosity of the displacing fluid may be increased or its effective permeability reduced until the value of M approaches unity or less. For instance, if the injected gas viscosity is increased, the gas mobility may be suppressed. Hence, the severity of the viscous fingering and the chance of developing premature breakthrough can be reduced, resulting in improved displacment efficiency. **Figure 6** demonstrates the effect of mobility ratio on the area sweep efficiency of an MGI process as reported by Habermann [40]. As can be seen from the figure, when M = 1, the ultimate areal sweep reaches as high as 99%, however, if M increases to 38.2, the areal sweep would decrease by more than 20%. The physical development of viscous fingers as the mobility ratio changes for the cases presented in **Figure 6** is demonstrated by the diagrams included in **Figure 7**. As can be seen in this figure, the displacement is characterised as stable if the value of M is one or lower. The effect of M as demonstrated through the above sweep values and **Figures 6** and 7 was for a homogeneous porous system. The presence of permeability heterogeneity would also make considerable contribution towards initiating and development of viscous fingering [41]. A high permeability layer would present a preferential flow path for the fingering of the injected gas causing early gas breakthrough and a low overall oil recovery factor [28, 29, 49].

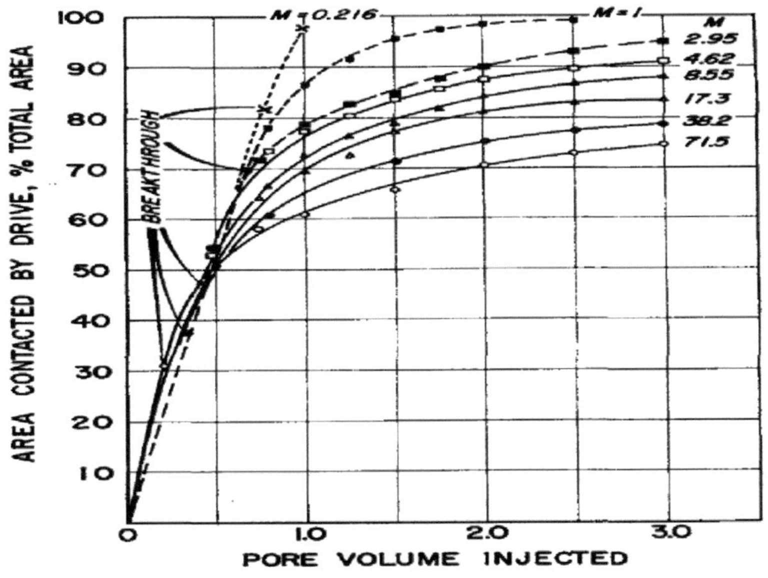

Figure 6.
Areal sweep efficiency as a function of mobility ratio and pore volumes of displacing phase injected for an MGI process [40].

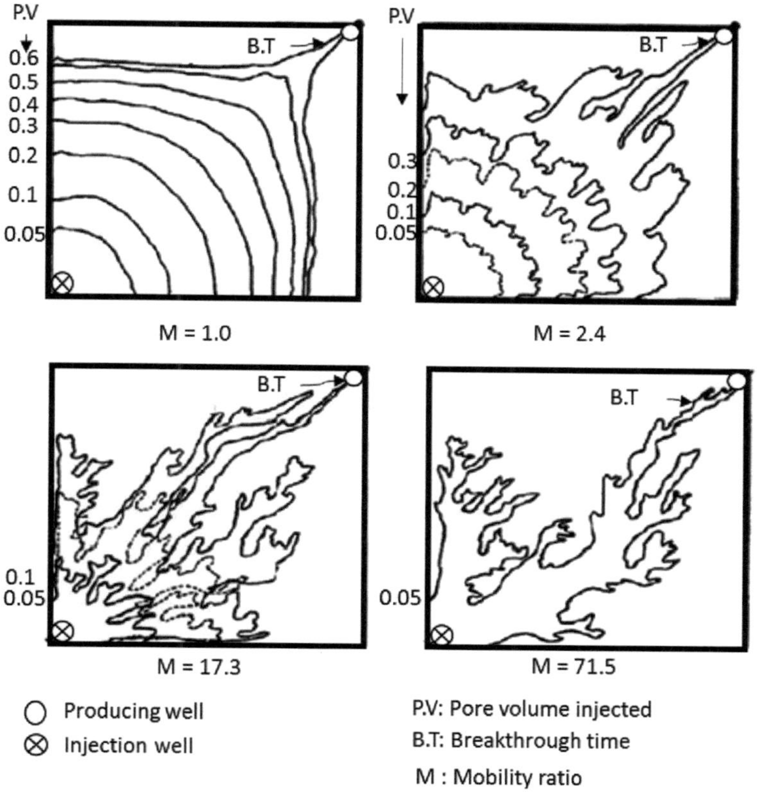

Figure 7.
Viscous fingering growth for different mobility ratio and injected pore volume [40].

3.2 Gravity segregation

As indicated earlier, another possible major technical challenge faced by an MGI process that influences the vertical sweep efficiency is the gravity segregation or gravity override. The injected gas (such as CO_2 or hydrocarbon gas) is usually less dense than the in-situ oil which may lead the injected gas to flow upwards, rather than lateral, forming a gravity tongue [49, 50]. Such a behaviour, similar to unfavourable mobility ratio, would result in early gas breakthrough and reduced vertical sweep efficiency in horizontal MGI processes as depicted in **Figure 8**. The effect of gravitational force on an MGI process has been studied by Moissis et al. [51] using numerical simulation. They found two dimensionless parameters of relevance, the dimensional density difference ($\Delta\rho$):

$$\Delta\rho = \frac{\rho_o - \rho_g}{\rho_g} \tag{4}$$

and the dimensionless gravity number (N_g):

$$N_g = \frac{\left(\rho_o - \rho_g\right) g K_e}{q \mu_o} \tag{5}$$

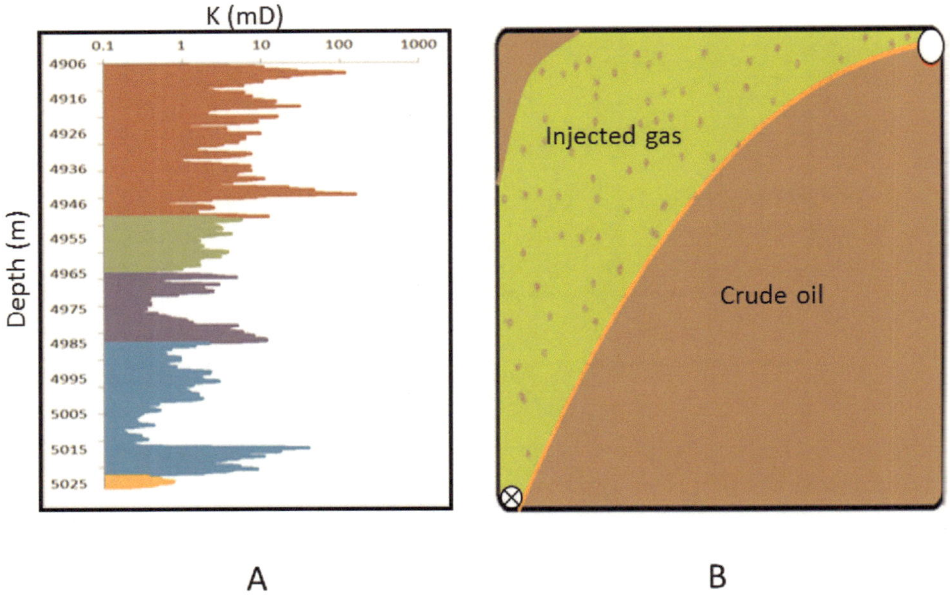

Figure 8.
(A) Reservoir heterogeneity due to permeability variation versus depth in field a located in South of Oman, (B) example effect of possible gravity segregation on vertical sweep efficiency.

where N_g represents the ratio of gravity forces to viscous forces, ρ_o and ρ_g are the oil and gas densities, respectively, K_e is equivalent permeability, μ_o is the oil viscosity, q is the flow rate of the less viscous fluid in the porous medium of interest, and g is the gravitational acceleration. The simulation results obtained by Moissis et al. [51] show that the gravity force does not influence viscous fingering growth at small N_g values indicating the dominance of the viscous forces under such a condition [51]. As N_g increases to larger values, the gravity force begins to influence the growth rate of viscous fingering in the upper part of the porous medium. For sufficiently large N_g values, gravity override completely dominates the displacement where, eventhough the viscous fingering can still occur near the gravity tongue, it is suppressed in the bottom part leaving this part of the porous medium completely unswept. Overall, as may be expected, with increase in N_g the gas breakthrough occurs earlier reducing the overall oil recovery [51].

Further interplay between the gravity and viscous forces towards controlling the efficiency of a gas flood may be deduced by further scrutiny of Eq. (5). Controlled by the magnitude of N_g, the effect of the gravitational force is expected to be even larger at high flood viscous ratios because the gravity to viscous forces ratio is inversely proportional to the viscosity of the fluid available in the porous medium. At the beginning of the flood, as defined by Eq. (5), this ratio is equal to N_g. However, as the displacement proceeds and more of the less viscous gas enters the porous medium at constant flow rate, the gravity to viscous forces ratio begins to increase resulting in more sever gravity override. Such an effect would be more pronounced in the case of floods characterised by a high viscosity ratio [51].

Scott [50] has suggested to combat the gravity segregation by adjusting the density of the miscible gas injected as part of an MGI. For example, the pressure within the formation can be maintained high enough so that the density of the injected fluid approaches that of the reservoir oil. However, for measurable outcomes in general, the density of the miscible fluid should be maintained within about 10% of the density of to be displaced in-situ oil [50]. Furthermore, this technique may be proven difficult and impractical if large injection volumes are required to maintain the reservoir pressure. Scott [50] has also indicated that the density adjustment may be obtained by injecting carbon dioxide or intermediate natural gas fractions (C_2H_6, C_3H_8 and C_4H_{10}). Carbon dioxide in its supercritical state is capable of exhibiting a density greater than that of the reservoir oil [50]. However, hydrocarbon gases alone may not normally achieve a density equal or close to that of the resident crude oil under typical reservoir conditions; therefore, sever gravity override could still occur. Another technique to increase the density of injected gas is the use of chemical additives; however, to date, suitable and viable chemical additives to be used for this purpose are yet to be developed [52]. As suggested in the literature, the mitigation of the gravity segregation can be possibly achieved by mobility or conformance control [52].

4. Gas mobility control techniques

As mentioned above, the major challenge with the ongoing MGI flooding in the oil field is the unfavourable mobility ratio. This challenge can be addressed by the implementation of several approaches as proposed in the literature (although mainly for CO_2 flooding) including water alternating gas flooding (WAG) [53–55], foam flooding [56–62] and increasing the gas viscosity using the addition of polymers as thickening agents [52, 63–71]. The common main objective of these approaches would be to control the gas mobility effectively and, as a result, increase the sweep efficiency of the gas flooding [72]. Further technical details about each of the abovementioned techniques are provided in the upcoming subsections of this chapter.

4.1 WAG process

As an EOR method, the WAG process is defined as the injection of a gas (e.g. CO_2 or hydrocarbon gases) and water alternately into an oil-bearing formation (**Figure 9**). The WAG injection scheme was initially proposed by Claudle and Dyes in 1958 [55] to improve sweep efficiency during gas flooding. Their study showed that this injection scheme would result in the reduction of the relative permeability to the gas phase and suppress its mobility. In other words, the WAG would improve the sweep efficiency of the injected gas by using water to control the gas mobility and stabilise the displacement front. In general, depending on the MMP of the in-situ oil, this technique can be classified into two categories of miscible and immiscible WAG displacements [73]; however, as reported in the literature, the majority (79%) of the historical WAG field applications fall into the miscible category [74, 75]. In some recent field applications, in an injection scheme similar to WAG, the produced gas has been reinjected through water injection wells to improve the oil recovery and help to provide pressure maintenance [76]. The majority of the WAG injection projects are found onshore (88%), and few others are reported to have been implemented in an offshore environment (12%) [75].

In general, there are a number of factors affecting the performance of the WAG process including the degree of reservoir heterogeneity, in-situ fluid properties, injection technique, miscibility conditions and other WAG parameters such as the individual gas and water slug sizes and their size ratio (WAG ratio), number of injection

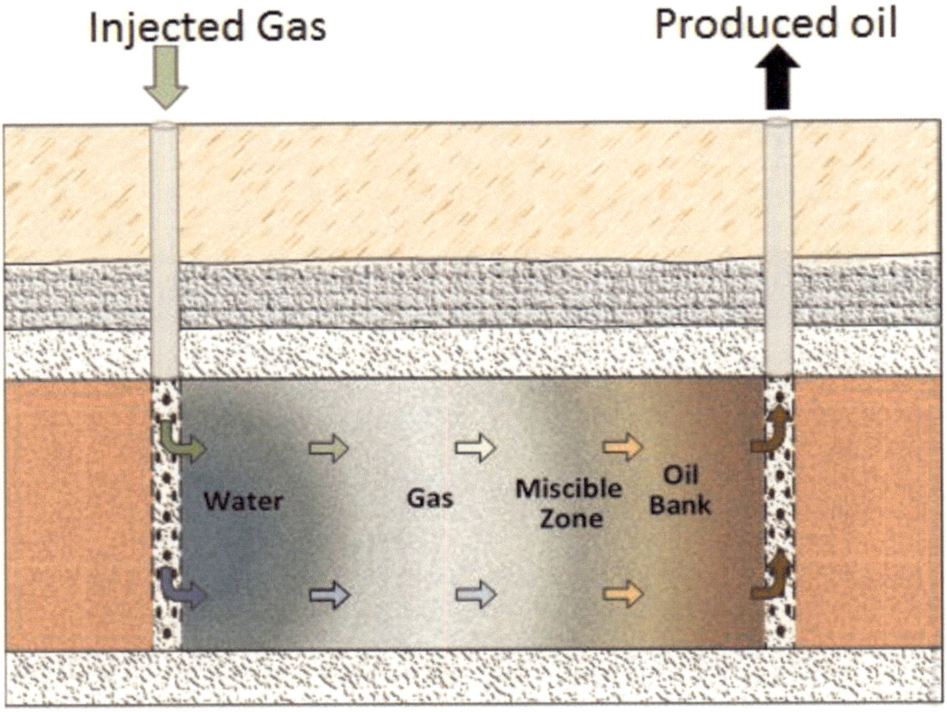

Figure 9.
A typical WAG injection process as an EOR method that involves the injection of gas and water alternatively into an oil reservoir.

cycles and injection rates [77–79]. Similar to other EOR processes, the WAG flooding has a number of advantages and disadvantages that will be presented and discussed below.

4.1.1 The mechanisms and factors influencing WAG flooding

During WAG injection, the improved recovery is not often achieved through modifying the fluid properties of each of the injected phases, it rather tends to combine the advantages of each of the continuous gas or continuous water floods through creating a synergism between the in-situ flow properties of the two phases if injected on their own. Overall, when WAG injection is applied in an oil reservoir, it may yield favourable outcomes through several mechanisms [80]. Firstly, the injection process may help to maintain the reservoir pressure above the MMP of the oil resulting the achievement of the more desirable miscible flood. Secondly, the injected gas mobility is reduced by supressing the gas relative permeability in any existing preferential flow channels. This is achieved by the increase in water saturation in these zones and therefore reduction in gas saturation suppressing the possibility of gas channelling and viscous fingering [52, 81, 82]. Thirdly, in the case of a miscible flood, the excellent microscopic displacement efficiency of the miscible gas flooding is put into use across a larger portion of the reservoir by the mobility control and conformance control provided by the water phase, leading to higher oil recovery. Lastly, compared with a continuous gas injection process (e.g. continuous MGI), the WAG flooding decreases the amount of the gas needed for injection leading to possible improvement in the economics of the overall flooding process. Considering the collective advantages mentioned here, the WAG injection process may become a viable option for some fields around the world.

Laboratory experiments have been used to study the effect of various parameters such as WAG slug size, WAG ratio (tapering), number of WAG cycle and injection flow rate on the performance of WAG [53, 54, 73, 83–89]. In general, these parameters show strong effects on the oil recovery trends of a WAG injection. It has been found that, in general, decreasing slug size and WAG ratio and increasing the number of WAG cycles would lead to a higher oil recovery [53, 80]. However, the optimum WAG ratio often depends on the wettability of reservoir rock, in-situ fluid properties and the type of gas being used as well as economic evaluations [53]. The optimum WAG ratio is considered as a key parameter for the successful implementation of a WAG injection process. A high WAG ratio may lead to an excessive water injection into the reservoir giving rise to the water blocking effect where the water phase would surround the trapped oil at low permeable zones and reduce accessibility by the injected gas decreasing the overall oil recovery. On the other hand, if the ratio is too low, the conformance control of the WAG flood would be lost and the injected gas would penetrate through the reservoir very fast under the effect of unfavourable mobility ratio and lead to early breakthrough. Overall, the experimental results have demonstrated that the WAG process may help to suppress viscous fingering and lead to increased oil recovery in gas flooding [74, 75].

4.1.2 Challenges of WAG flooding

The WAG injection has been successfully applied in several oilfields worldwide demonstrating that it could result in considerable incremental oil recovery at the field scale (5–10% of oil initially in place (OIIP)) [90]. However, some published literature also indicates that some of the field-scale WAG processes have not reached their expected target recovery factors, especially in naturally fractured, highly permeable and highly heterogeneous reservoirs [75]. Furthermore, the field-scale implementation of this technique has also helped to identify a number of challenges that may be faced by the field operators. Such challenges are presented and discussed below by first dividing them into the two categories of operational challenges versus those of subsurface reservoir related.

4.1.2.1 Operational issues

A numbers of operational related issues have been reported in the literature including [74, 75, 90].

4.1.2.2 Reduced injectivity

The ability to inject the required amounts of gas and water through the injection wells is critical towards achieving the desirable WAG performance. Reduced injectivity can result in a pressure reduction in the reservoir, which may impact on, for example, miscibility, performance of the displacement and the eventual production yield. This issue may be caused by changes in the phase relative permeabilities and/or near wellbore formation damage. In general, the field trials have shown that the reduced injectivity may be experienced for the water injection rather than the gas injection stage during the alternating injection of the two phases [75, 90].

4.1.2.3 Corrosion

Corrosion problems have been reported in many projects that have involved WAG injection. Often such issues have been encountered because the pre-existing injection and production facilities were not initially designed to handle the WAG injection process. Six fields are reported to have experienced corrosion problems, mainly on the injection facilities. The existing case studies indicate that in most cases, such problems could be adequately addressed by using corrosion-resistant materials in the manufacture of equipment, coating the flow-lines and chemical treatments [75, 90].

4.1.2.4 Asphaltene, scale and hydrate formation

Asphaltene and scale precipitation and hydrate formation are among other problems that have been experienced in various WAG field trials. These problems would lead to production disturbance and even flowlines blockage which may increase the operating costs of a WAG process. Three fields (East Vacuum, Wertz Tensleep, Mitsue) have experienced asphaltene precipitation, and two fields (Ekofisk and Wasson Denver) have reported the formation of hydrate in the injection wells due to the low temperature in the injectors or cold weather at the wellhead. Some of these problems could be resolved by chemical treatments [75, 90].

4.1.2.5 Subsurface reservoir issues

Besides the operational problems discussed above, there are also a number of issues related specifically to the subsurface and fluid flow in the bulk of the reservoir presenting challenges for the WAG implementation:

4.1.2.6 Premature gas breakthrough

Unexpected early gas breakthrough has been reported in several WAG field applications despite the fact that WAG is often implemented to combat this issue in particular. The main cause for this problem has often been inadequate characterisation of the reservoir, poor design of the WAG process or limitations imposed by the existing versus required infrastructure (e.g. limited number of injection/production wells). Regardless of the cause, early gas breakthrough would often occur due to gas channelling through highly permeable layers or gravity override [91, 92]. The early gas breakthrough leads to loss of reservoir pressure and lost miscibility in a miscible WAG project [93, 94]. As reported in the literature, five oil fields (University Block 9, Juravlevsko-Stepanovskoye, Lick Creek, Caroline and Snorre) have experienced this problem because of gas channelling [93, 95–98]. Unfortunately, this problem is hard to resolve as once occurred, its root causes (as mentioned at the beginning of this paragraph) are difficult to address. However, adequate reservoir characterisation before the implementation of this mobility control technique can be helpful in avoiding unexpected early gas breakthrough [75].

4.1.2.7 Oil trapping

Several studies have demonstrated the occurence of oil trapping by water in the WAG flooding [99–102]. This phenomenon is also referred to as water blocking [102]. During the WAG injection, the injected mobile water traps/encases the residual oil which then becomes difficult for the gas phase to access and mobilise. Therefore, a high residual oil saturation may be left behind in the reservoir even after WAG flooding. It has been determined that rock wettability and WAG ratio can strongly affect the oil trapping with being more sever in the case of water-wet rock formations or high WAG ratios [80, 99, 100, 103].

4.1.2.8 High water production

The injection of large amounts of the water into the reservoir (i.e. high WAG ratio) can cause high water saturation [104] leading to excessive water production and, hence, reduced oil recovery [105]. In addition, the excessive water production would require additional water treatment capacity that brings about additional costs impacting on the project economics [103].

4.2 Gas foam flooding process

4.2.1 Gas-foam generation and foaming agents

Gas-foam injection is another approach to combat the conformance and mobility limitations encountered in an MGI process. Furthermore, this technique may also bring about some of the advantages of the chemical EOR due to the chemical additives required for foam stabilisation and generally better foam generation. The foam flooding was first introduced by Bond and Halbrook in 1958 to show that the foam generated by the injection of an aqueous surfactant solution and miscible/immiscible gas could increase sweep efficiency [106]. With the favourable results obtained from the above study in the subsequent years, it was proposed to use foam injection as a means of gas mobility control. However, the concept did not become widely known and immediately adopted due to the lack of understanding of mobility control mechanisms behind the foam flooding [107].

In the context of fluid flow in porous media, a foam is generally defined as a gas-liquid mixture where the liquid phase exists as a continuous wetting phase in the rock, whereas all or parts of the gas form the discontinuous phase surrounded by a thin liquid film or Lamellae [60]. According to the literature, the research conducted in the area of gas foam flooding mostly relates to CO_2-EOR because the required chemicals are much easier to dissolve in CO_2 towards the generation of a CO_2 foam at reservoir conditions [72]. A gas foam may be stabilised by the addition of effective surfactants, which contain a hydrophobic and hydrophilic segment [72]. Surfactants then can be either water-soluble or CO_2-soluble [60, 108, 109]. The selection of surfactant depends on the reservoir conditions. If the reservoir condition is suitable for a surfactant to be soluble in the injected gas, then injection of water with the surfactant can be eliminated [110]. Numerous CO_2 soluble surfactants have been experimentally identified [56, 109, 110]. For example, the hydrocarbon-based ethoxylates surfactant has been suggested by Scheievelbein et al. as a CO_2 foam agent instead of using a water-soluble surfactant [110]. The other reported surfactant products include Tergitol TMN-6, oligo (vinyl acetate), poly(ethylene glycol) 2,6,8-trimethyl-4-nonyl ethers, and ethoxylated amine surfactant [111–115]. For miscible hydrocarbon gas flooding, only water-soluble surfactants can be used as the foaming agent because no effective surfactant directly soluble in hydrocarbon gases for gas-foam generation has been reported in the literature [57]. Nine water-soluble surfactants have been identified for foam generation with hydrocarbon solvents, including alkanolamides, amine oxides, betaine derivatives, ethoxylated and propoxylated alcohols and alkylphenols, ethoxylated and propoxylated fatty acids, ethoxylated fatty amines, fatty acid esters, fluorocarbon-based surfactants and sulfate and sulfonate derivatives [57]. As the temperature increases, most of the water-soluble surfactants become less soluble in water. Therefore, it may be necessary to evaluate the surfactant solubility in either CO_2 or water for application in high-temperature reservoirs [72, 116].

The foam used for gas foam flooding may be generated in several ways as discussed in the literature. It may be formed within the target porous media by alternating injection or co-injection of a suitable surfactant and gas (CO_2 or hydrocarbon gas mixture). In the case of CO_2 foam flooding, the foam can be formed when a surfactant is dissolved into CO_2 (usually in supercritical state) and then injected into the porous media, without requiring the injection of a liquid slug [59]. The foam can also be generated at the wellhead by the simultaneous injection of the gas and surfactant solution. Then, as the foam leaves the wellbore, it could be re-formed and strengthened as it enters the micropores of the reservoir rock [72].

As a gas foam enters a rock formation, it would need to propagate through the entire formation suppressing the high gas mobility for the whole duration of the flood. However, the injected foam is not often thermodynamically stable under in-situ conditions, and therefore, the two-phase foam system may collapse with time. On the other hand, as mentioned earlier, the passage of the fluids through the porous rock formation could result in the regeneration of the foam due to shearing effects applied by the micron-sized tortuous pores and pore channels [117]. Therefore, in order to have an effective foam for mobility control, the rate of in-situ foam generation would need to be equal to or greater than the rate of its decay [72]. In general, the foam propagation at the large reservoir scale and the foam stability are the main challenges faced by the gas foam flooding technique.

4.2.2 Main mechanisms of gas-foam flooding

A gas foam may be used as part of an EOR scheme for two purposes [57]. Firstly, it can be designed to reduce the gas mobility to a level that is comparable to or even less than that of the displaced oil so that the gas viscous fingering and channelling can be effectively suppressed. Thereby the areal sweep efficiency could be improved considerably. However, it is worth noting that the reduction level in the foam mobility has to be optimised and controlled to avoid the prohibitive pressure drop in the reservoir caused by extremely low foam mobility. Therefore as a compromise, a weak and modest foam may be generated by varying the surfactant concentration in a gas-foam injection [118]. The second possible purpose of using a gas foam is for conformance control or blocking of a thief gas channel to divert the injection fluids away from it and into other unswept lower permeability oil-rich zones to mobilise the otherwise bypassed oil [72, 109]. Typically, this can be achieved by the alternating injection of an aqueous solution with a high concentration of a surfactant [57]. The high concentration of surfactant then generates a strong foam that would flow in the highly permeable or thief zone [118] resulting in the diversion of the gas flow into the lower permeability zones.

The enhanced recovery of a gas foam injection is usually achieved through a number of different mechanisms as summarised and briefly discussed below.

4.2.2.1 Stabilising the displacement front

The efficiency of a fluid-fluid displacement in a porous medium is in general controlled by the three gravity, viscous and capillary forces [60, 119]. Therefore, the manipulation of these forces can result in enhanced recovery. Concerning the application of a gas foam during a gas flooding process such as MGI, the mobility control and, therefore, stabilisation of the flood front may be achieved by the higher viscosity and reduced relative permeability of the gas foam both relative to the case of injecting the gas on its own. Typically, these effects may be achieved through two mechanisms [60]. The first mechanism is related to the movment and re-arrangement of bubbles due to the local gradient in the surfactant concentration and, therefore, the interfacial tension. The surfactant movement within the liquid film (Lamellae) lowers the surface tension between the two phases (liquid and gas) that slows down the bubble motion and causes an increase in the gas phase effective viscosity [120–122]. The second mechanism that reduces the gas-foam mobility is gas trapping [123, 124]. As the foam injected and/or formed in a porous medium, as also indicated earlier, it prefers to flow through highly permeable and porous zones, while the low permeability areas with small pores remain occupied by the wetting phase [125] (**Figure 10**). Thus, the gas bubbles may enter and become trapped in the intermediate size pores, where a large fraction of foam bubbles are immobilised due to the high enough capillary pressure [59]. Nguyen et al. [126] found that the amount of trapped gas in this form is governed by several factors, such as the foam texture, pore geometry and pressure gradients. The blocked intermediate size pores decrease the pore volume available for the gas foam to flow through, thus the reduced relative permeability and suppressed gas-foam mobility [60].

A gas foam can help to combat gravity segregation too [60]. **Figure 11** demonstrates the effectiveness of a CO_2 foam towards stabilising the displacement front in the X-ray CT scanned core-flooding experiments conducted by Wellington and Vinegar [127]. As can be seen from the left-hand side images, the researchers found that CO_2 injection alone would lead to the formation of a gravity tongue, whereas the

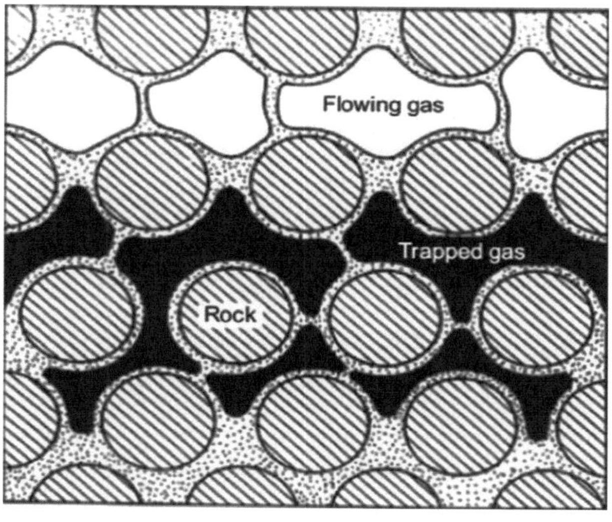

Figure 10.
A micro-pore illustration of foam flow and gas trapping in the porous media. The cross-hatched spaces represent the solid grains, and the dotted spaces indicate the wetting liquid [60, 117].

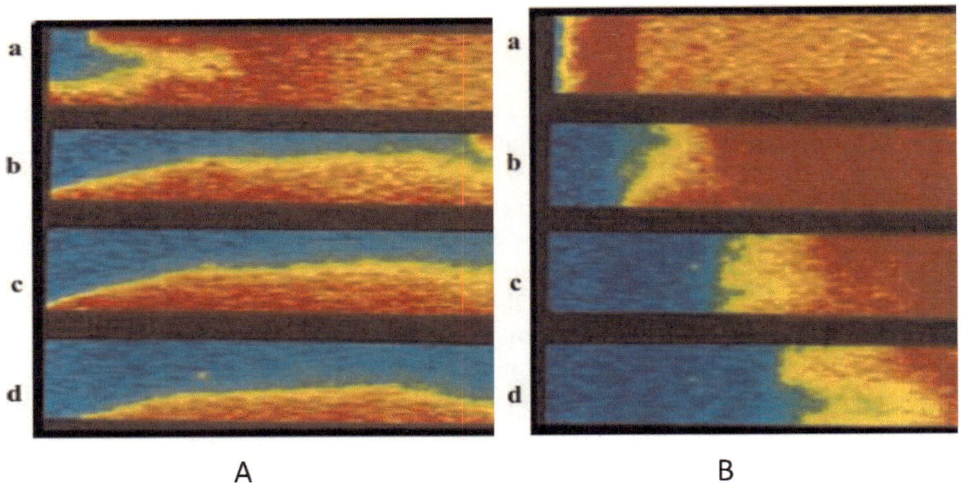

Figure 11.
X-ray CT scan images for (A) a CO_2 miscible flood (blue) in a core saturated with oil (red) and residual brine (yellow) and (B) CO_2-foam flooding (blue) in a core saturated with oil (red) and a surfactant solution (yellow) [72, 127].

right-hand side images show that the CO_2-foam injection prevented the gravity and viscous instabilities towards the uniform displacement of the in-situ oil.

Overall, based on the discussion presented so far, a gas foam would not change the gas phase density but exhibit its effectiveness by suppressing the gravity and viscous forces, leading to stabilisation of the displacement front.

4.2.2.2 Reducing the capillary force

Capillary pressure is usually held responsible for the bulk of the entrapped oil (often non-wetting phase) in rock formations. That is why Zhang et al. [128] point out that the removal of the trapped crude from a reservoir rock needs ultra-low interfacial tension through an emulsification mechanism. The capillary number as set out in Eq. (6) defines the ratio between viscous and capillary forces acting on a displacement. The lower the interfacial tension (low capillary forces), the higher the capillary number and, therefore, the more dominant would be the viscous forces resulting in higher recoveries.

$$N_c = \frac{K\Delta P}{\sigma L \cos\theta} \tag{6}$$

Where N_c: capillary number, dimensionless, K: absolute permeability of the porous medium, ΔP: pressure drop along the porous medium, σ: the interfacial tension between the two fluids, L: length of the porous medium, and θ: contact angle.

Once during foam injection, the surfactant in the injected slugs proceeds through the porous rock, different interactions occur at oil, foam and rock interface [129] leading to ultra-reduction of the interfacial tension between the oil and water resulting in the formation of an oil-in-water emulsion. Accordingly, the capillary force reduces to near zero allowing the emulsion to move through the pore throats (**Figure 12**) resulting in enhanced recovery [60].

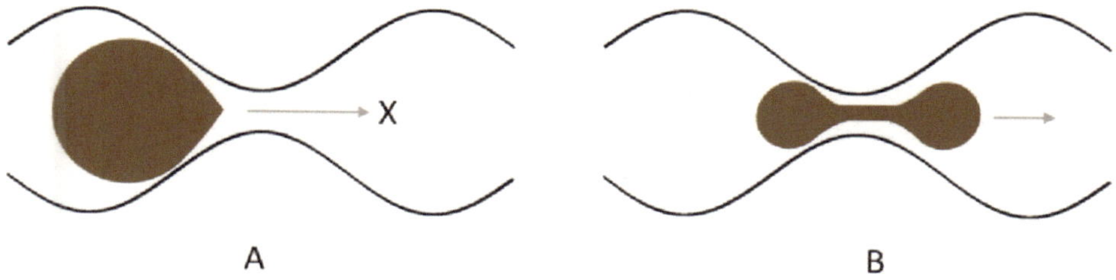

Figure 12.
(A) A high interfacial tension results in large capillary force, which prevents an oil droplet from crossing through the downstream pore throat, (B) ultra-low interfacial tension leads to near zero capillary force, which allows the oil droplet to flow through the pore throat and be produced [59].

4.2.2.3 Altering the rock wettability

The wettability of a porous rock formation is an essential factor to be taken into account in its characterisation because of its impact on the bond between oil and rock, the multiphase flow behaviour and distribution of fluid saturations in the reservoir [108]. Wettability alteration may occur in the foam flooding process due to the interactions between the surfactants used and the rock surface [59]. According to Eq. (6), the capillary number can also be increased by changing the contact angle, which means altering the rock wettability. As mentioned before, increasing the capillary number can result in lower residually trapped oil [59]. The importance of wettability alteration is not often considered in both experimental and simulation work, because of the erroneous assumption that all rocks remain water-wet during foam injection, and it is difficult to quantify the reservoir wettability in a meaningful and repeatable manner [130]. Although Charanjit and Bernard [131] do not agree that wettability may change due to a foaming agent, in a number of other studies, wettability alteration due the surfactant adsorption has been reported to change porous rocks from oil-wet to water-wet [131–133].

Overall, the foam injection process can enhance the oil recovery by mobility control in combination with ultra-low IFT and possible alteration of the rock wettability due to the presence of surfactant in the foam.

4.2.3 Challenges and field application for gas-foam EOR

The application of the gas-foam process in oil fields for mobility control has shown to be technically and economically challenging. This is because the effectiveness of a gas foam flooding highly depends on several parameters such as oil type, oil and water saturation, brine salinity and pH, surfactant formulation and concentration, reservoir heterogeneity, capillary pressure and gas flow rate [134, 135]. For example, a high oil saturation and low water saturation in the presence of light oil may cause the foam to decay and collapse [136]. As a consequence, before applying a foam EOR process, it is extremely important to gain a comprehensive understanding of the physical aspects of the process and how the foam may flow and behave once injected through a porous rock formation. The two main broad technical and operational difficulties in applying foam EOR at the large field scale are described below.

4.2.3.1 Foam stability and propagation

According to the numerous studies conducted to date, it may be difficult to achieve a stable and reliable foam generation under the harsh reservoir condition (high temperature and high salinity) often encountered and also control the propagation of the foam over large distances in the reservoir scale. Under high salinity and high temperature, the gas foam cannot be stabilised with the surfactant, because under such conditions the surfactant solubility in water or CO_2 would be reduced resulting in its precipitation onto the rock surface [115, 137]. In addition, with the loss of the surfactant, the necessary ultra-low IFT may not be achievable [138, 139]. The levels of oil and water saturations are other parameters that affect foam stability. Mayberry and Kam [140] examined the foam strength at different oil and water saturations. Their experimental results indicate that the apparent foam viscosity is significantly reduced at oil saturations greater or lower than a critical oil saturation. The presence of the oil in the formation has a strong effect on the foam rupture and breakdown due to the interactions occurring between the foam lamellae and the oil phase [141]. Law et al. [142] also found that foam is degraded if the oil saturation exceeds critical foaming oil saturation of the surfactant. It is also shown that the light and less viscous oils are more destructive to foam stability than heavy oils [135]. Moreover, the reservoir water saturation is crucial for the foam stability. When a foam is injected at water saturations below a critical value, which corresponds to a limiting capillary pressure, the foam may begin to coalesce and dry out. It should be noted that below the critical water saturation and above the critical oil saturation, the foam is eliminated [56, 136].

4.2.3.2 Scale-up from pilot to full field application

There have been several CO_2-foam trials performed since 1990 mainly in the United States [143–145]. Some of these, such as that performed in Joffre Viking oil field, were unsuccessful, because of the foam propagation control failure [146]. On the other hand, a few of the pilot tests have been successful, including that conducted in the Rock Creek Field [147] and Northward-Estes Field. In Northward-Estes Field, it was observed that the foam injection led to reduced CO_2 injectivity by 40–85% [143]. Several other pilot studies were conducted using CO_2 foam in East Vacuum Grayberg/San Andreas Unit [148] and SACROC Field in West Texas [149, 150], all of which proved that CO_2 mobility could be reduced and oil production increased. However, a transition from pilot scale to a wider field application has not been implemented due to various challenges such as issues associated with chemical supply and transportation, processing and separation of the produced fluids, offshore supply and also safety concern [151–154].

4.3 Direct gas thickeners

The use of direct gas thickeners is another method that brings together the combined possible advantages of using chemical additives and MGI. This technique has been recognised as a "game-changing technology" for mobility control, which was first reported in late 1960 [68, 69, 72, 155]. Since then, the interest in synthesising and designing affordable gas thickeners has been carrying on steadily. However, until now the term "gas thickener" has been used in laboratory investigations only, and its effectiveness has not yet been verified in any field-scale applications around the world. In general, this technique involves increasing the injected gas viscosity by directly adding chemicals that exhibit good solubility in common supercritical fluids (SCF) used for EOR such as CO_2 or hydrocarbon solvents. Chemicals that may increase the viscosity of an SCF include entrainers, conventional oligomers and polymers and small associating compounds [156]. In an ideal situation, chemical compounds need to be readily soluble in the dense CO_2 or hydrocarbons solvents and insoluble in both crude oil and brine at reservoir conditions [52]. It should be noted that the thickening level of the gas is not expected to affect its injectivity because this solution would exhibit a shear-thinning behaviour near the wellbore which facilitates the mobility of the thickened gas in this area but, the mobility ratio of the gas flood would be improved in the bulk of the rock formation leading to enhanced recovery (**Figure 13**). In addition, the thickened gas would uniformly flow into different zones, allowing the gas to also mobilise the trapped oil in the low permeable zones. In other words, this technique can be applied as a way of improving the flood conformance and mobility control as illustrated in **Figure 13**.

Two fundamental strategies have been introduced in the literature to increase the injected gas viscosity [157].

Direct dissolution of polymers: In this strategy, a gas thickener is typically a synthesised or identified polymer or oligomer that promotes attractive interactions and dissolution with gas molecules. However, it has been recognised that the use of polymers with extraordinary molecular weight for the above purpose would be quite challenging since most of the SCF fluids are very stable and weak solvents due to the very low dielectric constant, no dipole momentum and sometimes low density. The

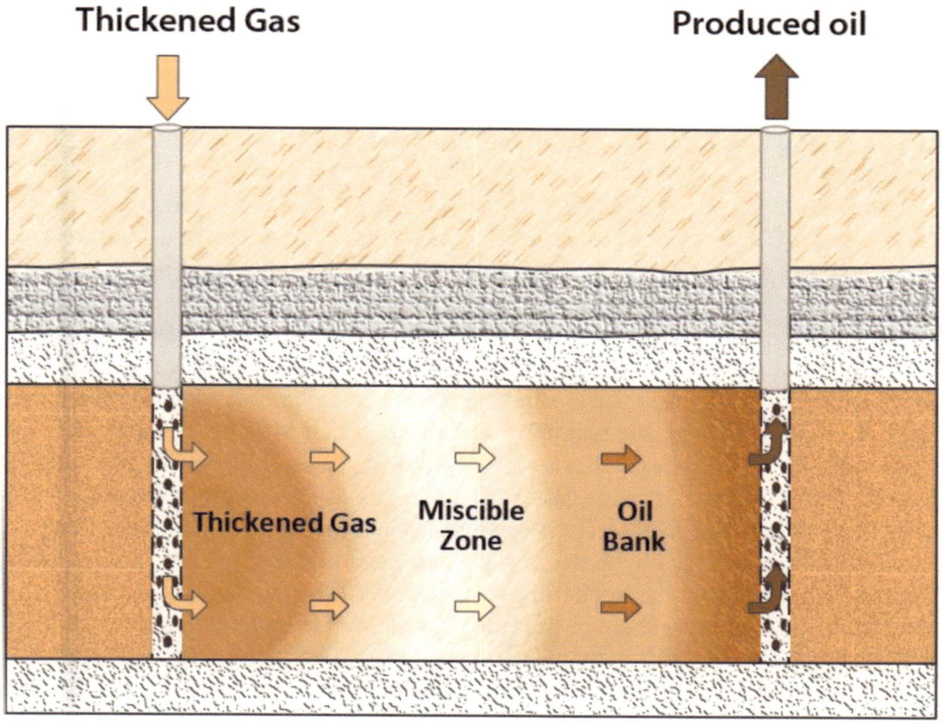

Figure 13.
Simplified illustration of a thickened gas flooding.

intermolecular attractions between the polymer molecules are typically strong enough at ambient temperature so that even stirring them would be insufficient to attain dissolution. Therefore, they may only dissolve in a gas solvent at elevated pressure and temperature because such conditions give rise to the intermolecular forces between the solvent-polymer segments or solvent-solvent or polymer segment-segment pairs in the solution given by difference on the free volume between the polymer and gas solvent and the free energy [158]. In addition, heat may be required to weaken intermolecular interactions between the polymer molecules (e.g. hydrogen bond) [158]. Another approach for obtaining high solubility of the polymer in solvents is to introduce associating or functional groups in the polymer's molecular chains, for example to become CO_2 philic, and therefore assist the polymer dissolution in the solvent [159, 160]. Some examples of the associated polymers include polyvinyl acetate (PVAc), oligo (3-acetoxy oxetane), poly [(1-O-(vinyloxy) ethyl-2, 3, 4, 6-tetra-O-acetyl-β-D-glucopyranoside)] and amorphous polylactic acid [161, 162]. Once the molecules of the polymer are dissolved in the solvent, the intermolecular/intramolecular association may occur which would result in an increased solution viscosity. Some of the polymers can increase the solvent viscosity significantly by simply changing the thickener concentration or by twining their molecular structure like a hair between different polymer chains [163].

Dissolution of small molecules (self-assembling and associating compound): The second strategy is focusing on the design of small-molecules material that contains a self-assembling and associating compound to form a viscosity-enhancing supramolecular network structure in the solution. Such a material contains an associating group composed of a solvent philic segment that facilitates dissolution and one or more solvent-phobic segments that would induce the intramolecular association with neighbouring molecules, thereby molecular association establishing a viscosity enhancement for the solution, but its impact on viscosity could be minimal [72]. The small-molecules thickeners have shown little success to thicken CO_2 and light alkane solvents primarily because these are regarded as weak solvents for the ionic and polar associating compounds that are commonly composed into the small-molecules thickeners [71, 157].

Overall, a polymeric or small-molecules compound thickener capable of dissolving into CO_2 or light hydrocarbon solvents has to be identified to increase the solution viscosity under typical field conditions. The ideal chemical additives are those that can effectively increase the viscosity of the injected gas very close to that of the crude oil. Furthermore, a viscosified gas used for EOR has to be transparent and single phase rather than opaque viscous solution in order to be capable of flowing through micro-pore throats in rock formations [157]. A viscosified gas with the above-described desirable characteristics used for an MGI process can suppress the gas mobility in the reservoir reducing the severity of viscous fingering and the chance of developing premature gas breakthrough and high production gas oil ratio (GOR). As a result, the sweep efficiency would be improved for the gas flood. Various studies conducted over the past several decades have resulted in successful laboratory-scale progress in thickening of CO_2 and NGL (natural gas liquefied). The successful CO_2 thickeners include the fluoroacrylate-styrene copolymer polyFAST and poly(dimethylsiloxane)-toluene solutions [160]. These two thickeners have been found to be capable of increasing the CO_2 viscosity by approximately 10 and 4 fold, respectively, at dilute concentrations [160]. A drag-reducing agent (DRA) poly(α-olefin) was presented as the most significant thickener that can increase the viscosity of the NGL [164].

4.3.1 Challenges and opportunity for gas thickeners

The use of gas thickeners has the potential to eliminate many of the earlier mentioned challenges and difficulties associated with WAG and gas-foam injections. However, the discovery of inexpensive polymers or small-molecules materials soluble in CO_2 or alkane solvents has so far been a major challenge. Furthermore, the performance of none of the identified or synthesised thickeners has been verified in even a field pilot test yet.

In general, the following challenges have hindered the identification of effective thickeners that could be used for a gas flood:

Thickener solubility: The attainment of adequate solubility has been the primary obstacle in finding viable thickeners because most of the designed and identified polymers for the CO_2 and hydrocarbon gases exhibit extremely low solubility unless a large volume of a co-solvent (e.g., 10–15 wt% toluene) is added. The reason behind this problem is that CO_2 and alkane gases are poor solvents for extremely high molecular weight, polar and ionic-associated groups that are composed in small-molecules thickeners. The alkane gases (methane and ethane) do not have dipole or quadruple moments, so the dispersion interactions are dominant with these solvents. Thereby, alkane gases would not be suitable SCF solvents unless the density of these solvents is increased considerably by increasing the system pressure. Unlike alkane gases, CO_2 has a substantial quadrupole moment that induces quadruple interaction as the temperature is low [158]. In addition, CO_2 acts as a Lewis acid for the polymers containing oxygen [165]. In general, as mentioned before, a polymeric CO_2 thickener needs to contain a CO_2-philic function group that facilitates the polymer solubility and CO_2-phobic function group that promotes intermolecular associations to enhance the viscosity [166]. To date, solubility remains a key major challenge in the identification of an inexpensive thickener for CO_2 and hydrocarbon solvents.

Cost and environmental persistence: The high price and environmental issues are other challenges that impede the use of the identified or developed thickeners to date in field applications. In fact, most of such thickeners are unaffordable and/or unavailable in large enough quantities. The requirment of an organic co-solvent to obtain the necessary dissolution levels further adds to the cost. Moreover, some of the developed thickeners, such as fluoroacrylate-styrene copolymers (polyFAST) and semi-fluorinated trialkyltin fluorides, are fluorinated compounds that contain Fluorine. These thickeners have been identified as the best thickeners for CO_2 and NGL, respectively. However, the fluorine in these thickeners would bring about potential negative effects on the environment making them unsuitable for EOR applications [52, 72, 157, 160].

5. Mobility control: feasibility evaluation field A

Overall, from the discussions presented so far, it is clear that each of the proposed mobility control methods, as applicable to an MGI process, has its own challenges and deficiencies. The possible field-scale implementation of each method often depends primarily on the in-situ conditions and specific characteristics of the field of interest. The objective of this section of the chapter is to present an evaluation of the applicability of each of the techniques discussed earlier in Field A given its specific conditions and characteristics.

5.1 WAG technique

As mentioned earlier, in field applications, the WAG process has been applied successfully in a number of oil fields around the world [74, 75]. A total of 72 field-scale miscible and immiscible WAG projects were reviewed by Skauge et al. that have utilised hydrocarbon or non-hydrocarbon gases. Majority of these projects have been successful resulting in incremental oil recoveries in the range of 5–10% of OIIP. For successful projects, the WAG process consistently yielded better oil recovery than that could be achieved with continuous gas injection even though, often, a large amount of oil (35–65% of OIIP) would still be left behind [52]. Some of reviewed projects have also been unsuccessful due to operational and/or reservoir related difficulties such gas gravity segregation, extreme reservoir heterogeneity, excessive water production, corrosion, scale and/or hydrate formation, etc. [74] In the case of Field A, in-situ water saturation is very low (<10%) and, therefore, the field surface facilities and well completions are not designed to inject or handle large amounts of water. Therefore, the WAG strategy is not the best choice to implement in this field.

5.1.1 Gas foam technique

It was previously discussed that the gas foam injection process has been tried at the pilot scale in some fields in the United States and Canada. However, this technique has never been performed in any field in the Middle East due to the difficulties of finding a suitable surfactant (water soluble) or due to the harsh reservoir conditions encountered including high salinity and high temperature. Although, there has been a number laboratory-scale studies done to date evaluating the application the technique under conditions encountered in this region. For example, in a recent study conducted by Sumaiti et al. [56, 111], the foamability and mobility of CO_2-ethoxylated amine in carbonate cores were investigated at a salinity of 220,000 ppm and temperature of 393 K. The foamability of Ethomeen (C_{12}) and apparent foam viscosity increase were confirmed at these conditions. In addition, CO_2-foam core flooding obtained 8.89% of additional oil recovery. However, the availability of CO_2 is very limited in the Middle East. Concerning Field A, the reservoir presents a harsh environment with a formation brine salinity of 275,000 ppm and a reservoir temperature of 377 K with low in-situ water saturation and a very light oil (42° API). It is extremely difficult to find a surfactant, especially water-soluble, which can work under these conditions. For the CO_2-foam process, there is a lack of adequate CO_2 availability in Oman. As a result, it is expected that achieving adequate foam stability would be a major challenge to implement a gas-foam process in Field A.

5.2 Direct thickened technique

As discussed earlier, several laboratory-scale studies have been conducted to date to find and/or develop direct thickeners for CO_2 and NGL. However, the cost and environmental issues associated with these thickeners have prevented their application beyond the laboratory scale [167]. As outlined earlier, this technique has several distinct advantages compared with the other two mobility/conformance control techniques of WAG and gas-foam injection. Firstly, a screened thickener additive would be thermodynamically stable and chemically inert (with no or minimal interaction with reservoir sediments), making it ideal for application in harsh reservoir conditions (i.e. high formation salinity and temperature). Secondly, the gas viscosity increase achievable by a thickener does not dependent on rock characteristics, properties and saturations of other fluids in the reservoir and injection flow rates. Thirdly, it eliminates the need for water co-injection which minimises the chance of excessive water production and treatment requirements substantially and eliminates the water blocking effect too. Lastly, it has been demonstrated at the laboratory scale that this technique can increase the sweep efficiency considerably because of delayed gas breakthrough and improved gas mobility. Hence, it is believed that CO_2 or AG mixture thickening may be the only viable technique for Field A to counteract unfavourable mobility conditions present in the Field and further enhance the oil recovery of the current ongoing MGI.

6. Conclusion and recommendations

This chapter presents the process of miscible gas injection (MGI) and the implementation of MGI in the petroleum industry especially for the recovery of light oil. It briefly discussed the challenges associated with the MGI flooding, and several solutions proposed in the literature to overcome these challenges include: water alternating gas flooding (WAG), foam flooding and the use of thickening agents. Despite many efforts made to date to identify a viable approach to counteract unfavourable mobility conditions and improve sweep efficiency. These approaches are not applicable in the fields as means of mobility control at field scale. Therefore, a further work requires that can improve the industry's confidence in employing these approaches at the field scale using numerical simulation followed by economic analysis to investigate and verify the feasibility of these techniques for field applications.

Author details

Nasser Mohammed Al Hinai[1]* and Ali Saeedi[2]

1 Petroleum Development Oman, Petroleum Engineering, Sultanate of Oman

2 Department of Mining Engineering, Curtin University, Australia

*Address all correspondence to: nasser.alhinai5@gmail.com

References

[1] Alvarado V, Manrique E. Enhanced oil recovery: An update review. Energies. 2010;**3**(9):1529-1575

[2] Muggeridge A, Cockin A, Webb K, Frampton H, Collins I, Moulds T, et al. Recovery rates, enhanced oil recovery and technological limits. Philosophical Transactions of the Royal Society of London A: Mathematical, Physical and Engineering Sciences. 2014;**372**(2006): 20120320

[3] Ahmed T. Reservoir Engineering Handbook. USA: Elsevier; 2006

[4] Lake LW. Enhanced Oil Recovery. Upper Saddle River, New Jersey: Prentice-Hall; 1989

[5] Nwidee LN, Theophilus S, Barifcani A, Sarmadivaleh M, Iglauer S. EOR processes, opportunities and technological advancements. Chemical Enhanced Oil Recovery (cEOR)—A Practical Overview. 2016:2-52

[6] Research GV. Enhanced Oil Recovery Market Size, Share & Trend Analysis Report by Technology, by Application and Segment Forecasts, 2018-2025. CISION PR Newswire: Grand View Research; 2018. p. 129

[7] Al-Mutairi SM, Kokal SL. EOR potential in the Middle East: Current and future trends. In: SPE EUROPEC/EAGE Annual Conference and Exhibition. Vienna, Austria: Society of Petroleum Engineers; 2011

[8] Al-Adawy M, Nandyal M. Status and scope for EOR development in Oman. In: Middle East Oil Show. Bahrain: Society of Petroleum Engineers; 1991

[9] Henni A. Oman champions EOR in the Middle East. Journal of Petroleum Technology. 2014;**66**(11):86-89

[10] Terry RE. Enhanced oil recovery. Encyclopedia of Physical Science and Technology. 2003;**18**:503-518

[11] Zick A. A combined condensing/ vaporizing mechanism in the displacement of oil by enriched gases. In: SPE Annual Technical Conference and Exhibition. New Orleans: Louisiana Society of Petroleum Engineers; 1986

[12] Asgarpour S. An overview of miscible flooding. Journal of Canadian Petroleum Technology. 1994;**33**(02):13-15

[13] Solutions MT. Miscible Gas Injection: Facts about Miscible Displacements. 2022. Available from: http://www.mktechsolutions.com/Miscible%20Gas%20Displacements.html

[14] Dong M, Huang S, Dyer SB, Mourits FM. A comparison of CO2 minimum miscibility pressure determinations for Weyburn crude oil. Journal of Petroleum Science and Engineering. 2001;**31**(1):13-22

[15] Stalkup FI. Displacement behavior of the condensing/vaporizing gas drive process. In: SPE Annual Technical Conference and Exhibition. Dallas, Texas: Society of Petroleum Engineers; 1987

[16] Holm L, Josendal V. Mechanisms of oil displacement by carbon dioxide. Journal of Petroleum Technology. 1974; **26**(12):1427-1438

[17] Simon R, Graue D. Generalized correlations for predicting solubility, swelling and viscosity behavior of CO2-crude oil systems. Journal of Petroleum Technology. 1965;**17**(01):102-106

[18] Gu Y, Hou P, Luo W. Effects of four important factors on the measured

minimum miscibility pressure and first-contact miscibility pressure. Journal of Chemical & Engineering Data. 2013; **58**(5):1361-1370

[19] Hemmati-Sarapardeh A, Ayatollahi S, Ghazanfari M-H, Masihi M. Experimental determination of interfacial tension and miscibility of the CO2–crude oil system; temperature, pressure, and composition effects. Journal of Chemical & Engineering Data. 2013;**59**(1):61-69

[20] Holm L. Miscible displacement. In: Petroleum Engineering Hand Book. Richardson, TX: Society of Petroleum Engineers; 1987. pp. 1-45

[21] Stalkup FI Jr. Status of miscible displacement. Journal of Petroleum Technology. 1983;**35**(04):815-826

[22] Belhaj H, Abukhalifeh H, Javid K. Miscible oil recovery utilizing N_2 and/or HC gases in CO_2 injection. Journal of Petroleum Science and Engineering. 2013;**111**:144-152

[23] Whittaker S, Perkins E. Technical Aspects of CO2 Enhanced Oil Recovery and Associated Carbon Storage. Melbourne: Global CCS Institute; 2013

[24] Algharaib MK. Potential applications of CO2-EOR in the Middle East. In: SPE Middle East Oil and Gas Show and Conference. Manama, Bahrain: Society of Petroleum Engineers; 2009

[25] Al-Hajeri SK, Negahban S, Bin-dhaaer GS, Al-Basry AH. Design and implementation of the first CO2-EOR pilot in Abu Dhabi, UAE. In: SPE EOR Conference at Oil & Gas West Asia. Muscat, Oman: Society of Petroleum Engineers; 2010

[26] Al-Hajeri SK, Ghedan SG, Al-basry AH, Negahban S. Integrated history matching simulation study of the first miscible CO2-EOR pilot in Abu Dhabi heterogeneous carbonate oil reservoir. In: SPE Middle East Oil and Gas Show and Conference. Manama, Bahrain: Society of Petroleum Engineers; 2011

[27] Saadawi HN. Surface facilities for a CO2-EOR project in Abu Dhabi. In: SPE EOR Conference at Oil & Gas West Asia. Muscat, Oman: Society of Petroleum Engineers; 2010

[28] Al-basry AH, Al-Hajeri SK, Saadawi HN, Aryani A, Mohamed F, Al Obeidi A, et al. Lessons learned from the first miscible CO2-EOR pilot project in heterogeneous carbonate oil reservoir in Abu Dhabi, UAE. In: SPE Middle East Oil and Gas Show and Conference. Manama, Bahrain: Society of Petroleum Engineers; 2011

[29] Alajmi AF, Algharaib M, Gharbi R, Alenezi H, AlMatar B, Al-Enezi K. Experimental investigation of CO2 miscible flood in West Kuwait. Journal of Engineering Research. 2015;**3**(3)

[30] Al Hadhrami A, Al-Jabri A, Al-Alawi HM, Proot M, Schulz RM, Barwani S, et al. Challenges and learning from developing deep high pressure sour reservoirs in Southern Oman by EOR MGI. In: Abu Dhabi International Petroleum Exhibition and Conference. UAE: Society of Petroleum Engineers; 2014

[31] Al-Hadhrami AK, Davis DW, Deinum G, Soek H. The design of the first miscible sour gasflood project in Oman. In: International Petroleum Technology Conference. Dubai, UAE: International Petroleum Technology Conference; 2007

[32] Hinai NMA, Saeedi A, Wood CD, and Valdez R. A Numerical Study of Using Polymers to Improve the Gas

Flooding in the Harweel Cluster. In SPE Reservoir Characterisation and Simulation Conference and Exhibition. Abu Dhabi. UAE Society of Petroleum: Engineers; 2017

[33] Al-Riyamy QM, Kelly SB, Al-Rawahi A, Matsuura T, Davis D, Henning RL, et al. Precambrian field Oman from Greenfield to EOR. In: SPE Middle East Oil and Gas Show and Conference. Kingdom of Bahrain Society of Petroleum Engineers; 2005

[34] Laws M, Al-Riyami A, Soek H, Abdulkadir R. Optimization of Stimulation Techniques for HP, Deep Wells in the Harweel Cluster, Oman in Paper SPE 93494 Presented in the SPE Middle East Oil and Gas Show and Conference; Kingdom of Bahrain: SPE; Mar. 2005

[35] Kokal S, Al-Kaabi A. Enhanced Oil Recovery: Challenges & Opportunities. London, U.K: World Petroleum Council: Official Publication; 2010. pp. 64-68

[36] Chierici GL. Principles of Petroleum Reservoir Engineering. Vol. 2. Germany: Springer Science & Business Media; 2012

[37] Nicolaides C, Jha B, Cueto-Felgueroso L, Juanes R. Impact of viscous fingering and permeability heterogeneity on fluid mixing in porous media. Water Resources Research. 2015; **51**(4):2634-2647

[38] Chang Y-B, Lim MT, Pope GA, Sepehrnoori K. CO2 flow patterns under multiphase Flow. Heterogeneous Field-Scale Conditions. 1994;**9**(03):208-216

[39] Homsy GM. Viscous fingering in porous-media. Annual Review of Fluid Mechanics. 1987;**19**(1):271-311

[40] Habermann B. The Efficiency of Miscible Displacement as a Function of Mobility Ratio. Pasadena, CA: Petroleum Transactions, AIME; 1960

[41] Araktingi UG, Orr FM Jr. Viscous Fingering in Heterogeneous Porous Media. SPE Reservoir Engineering. 1993; **9**(03)

[42] Fu X, Cueto-Felgueroso L, Juanes R. Viscous fingering with partially miscible fluids. Physical Review Fluids. 2017; **2**(10):104001

[43] Moortgat J. Viscous and gravitational fingering in multiphase compositional and compressible flow. Advances in Water Resources. 2016;**89**: 53-66

[44] Suekane T, Ono J, Hyodo A, Nagatsu Y. Three-dimensional viscous fingering of miscible fluids in porous media. Physical Review Fluids. 2017; **2**(10):103902

[45] Brock DC, Orr FM Jr. Flow visualization of viscous fingering in heterogeneous porous media. In: SPE Annual Technical Conference and Exhibition, Dallas. Texas: Society of Petroleum Engineers; 1991

[46] Blunt M, Christie M. Theory of viscous fingering in two phase, three component flow. SPE Advanced Technology Series. 1994;**2**(02):52-60

[47] Cheek RE, Menzie DE. Fluid Mapper MODEL Studies of Mobility Ratio. New York: Society of Petroleum Engineers; 1955

[48] Sahimi M, Rasaei MR, Haghighi M. Gas injection and fingering in porous media. In: Gas Transport in Porous Media. Springer; 2006. pp. 133-168

[49] Pande K. Effects of gravity and viscous crossflow on hydrocarbon miscible flood performance in

heterogeneous reservoirs. In: SPE Annual Technical Conference and Exhibition. Washington, D.C.: Society of Petroleum Engineers; 1992

[50] Scott J. Control of gravity segregation by high density fluid injection. US3661208A. 1972

[51] Moissis D, Wheeler M, Miller C. Simulation of miscible viscous fingering using a modified method of characteristics: Effects of gravity and heterogeneity. SPE Advanced Technology Series. 1993;**1**(01): 62-70

[52] Enick RM, Olsen DK, Ammer JR, Schuller W. Mobility and conformance control for CO2 EOR via thickeners, foams, and gels—A literature review of 40 years of research and pilot tests. In: SPE Improved Oil Recovery Symposium. Tulsa, Oklahoma, USA: Society of Petroleum Engineers; 2012

[53] Bhatia J, Srivastava J, Sharma A, Sangwai JS. Production performance of water alternate gas injection techniques for enhanced oil recovery: Effect of WAG ratio, number of WAG cycles and the type of injection gas. International Journal of Oil, Gas and Coal Technology. 2014;**7**(2):132-151

[54] Bhatia JC, Srivastava J, Sangwai JS, Sharma A. Investigations on gas trapping phenomena for different EOR-water alternate gas injection methodologies. In: International Petroleum Technology Conference. Bangkok, Thailand: International Petroleum Technology Conference; 2011

[55] Caudle BH, Dyes AB. Improving Miscible Displacement by Gas-Water Injection. Dallas, Texas: Society of Petroleum Engineers; 1958

[56] Al Sumaiti A, Shaik AR, Mathew ES, Al Ameri W. Tuning foam parameters for mobility control using CO2 foam: Field application to maximize oil recovery from a high temperature high salinity layered carbonate reservoir. Energy & Fuels. 2017;**31**(5):4637-4654

[57] Karin M, Novosad JJ. Adsorption of foam-forming surfactants for hydrocarbon-miscible flooding at high salinities. In foams: Fundamentals and applications in the petroleum industry. Advances in Chemistry. 1994;**242**:259-316

[58] Schramm LL, Mannhardt K, Novosad JJ. Selection of oil-tolerant foams for hydrocarbon miscible gas flooding. In: Proceedings 14th International Workshop and Symposium. Salzburg: Int. Energy Agency Collab. Proj. on EOR; 1993

[59] Talebian SH, Masoudi R, Tan IM, Zitha PL. Foam assisted CO2-EOR; concepts, challenges and applications. In: SPE Enhanced Oil Recovery Conference. Kuala Lumpur, Malaysia: Society of Petroleum Engineers; 2013

[60] Talebian SH, Masoudi R, Tan IM, Zitha PLJ. Foam assisted CO2-EOR: A review of concept, challenges, and future prospects. Journal of Petroleum Science and Engineering. 2014;**120**:202-215

[61] Xu X, Saeedi A, Liu K. Laboratory studies on CO 2 foam flooding enhanced by a novel amphiphilic ter-polymer. Journal of Petroleum Science and Engineering. 2016;**138**:153-159

[62] Xu X, Saeedi A, Zhang Y, Liu K. Laboratory investigation on CO2 foam flooding for mature fields in Western Australia. In: Offshore Technology Conference Asia. Kuala Lumpur, Malaysia: Offshore Technology Conference; 2016

[63] Bae J, Irani C. A laboratory investigation of viscosified CO2 process.

SPE Advanced Technology Series. 1993; **1**(01):166-171

[64] Dandge D, Heller J. Polymers for mobility control in CO2 floods. In: SPE International Symposium on Oilfield Chemistry. San Antonio: Texas Society of Petroleum Engineers; 1987

[65] Enick R. The effect of hydroxy aluminum disoaps on the viscosity of light alkanes and carbon dioxide. In: SPE International Symposium on Oilfield Chemistry. Anaheim: California Society of Petroleum Engineers; 1991

[66] Heller J, Dandge D, Card R, Donaruma L. Direct thickeners for mobility control of CO2 floods. Society of Petroleum Engineers Journal. 1985; **25**(05):679-686

[67] Terry RE, Zaid A, Angelos C, Whitman DL. Polymerization in supercritical CO2 to improve CO2/oil mobility ratios. In: SPE International Symposium on Oilfield Chemistry. San Antonio: Texas Society of Petroleum Engineers; 1987

[68] C Jr AR. Oil recovery process by miscible displacement. U.S. Patent 3,444,930 May 20. 1969

[69] Dauben DL, Reed J, Shelton J, Lyman Y. Recovery of oil with viscous propane. U.S. Patent 3,570,601. 1971

[70] Dhuwe A, Lee J, Cummings S, Beckman E, Enick R. Small associative molecule thickeners for ethane, propane and butane. The Journal of Supercritical Fluids. 2016;**114**:9-17

[71] Lee J, Dhuwe A, Cummings SD, Beckman EJ, Enick RM, Doherty M, et al. Polymeric and small molecule thickeners for CO2, ethane, propane and butane for improved mobility control. In: SPE Improved Oil Recovery Conference. Tulsa, Oklahoma: USA Society of Petroleum Engineers; 2016

[72] Enick R, Olsen D. Mobility and conformance control for carbon dioxide enhanced oil recovery (CO2-EOR) via thickeners, foams, and gels—A detailed literature review of 40 years of research. Contract DE-FE0004003. Activity. 2012;**4003**(01)

[73] Kulkarni MM, Rao DN. Experimental investigation of miscible and immiscible water-alternating-gas (WAG) process performance. Journal of Petroleum Science and Engineering. 2005;**48**(1-2):1-20

[74] Skauge A, Stensen J. Review of WAG field experience. In: Oil Recovery–2003, 1st International Conference and Exhibition, Modern Challenges in Oil Recovery. Moscow, Russia: Gubkin University; 2003

[75] Christensen JR, Stenby EH, Skauge A. Review of WAG field experience. SPE Reservoir Evaluation & Engineering. 2001;**4**(02)

[76] Masoud M. Comparing carbon dioxide injection in enhanced oil recovery with other methods. Austin Chemical Engineering. 2015;**2**(2): 1019

[77] Surguchev LM, Korbol R, Haugen S, Krakstad OS. Screening of WAG injection strategies for heterogeneous reservoirs. In: European Petroleum Conference. Cannes, France: Society of Petroleum Engineers; 1992

[78] Chen S-M, Allard DR, Anli J. Factors affecting solvent slug size requirements in hydrocarbon miscible flooding. In: SPE Enhanced Oil Recovery Symposium. Tulsa, Oklahoma: Society of Petroleum Engineers; 1984

[79] Sanchez NL. Management of water alternating gas (WAG) injection projects. In: Latin American and Caribbean Petroleum Engineering Conference. Caracas, Venezuela: Society of Petroleum Engineers; 1999

[80] Han L, Gu Y. Optimization of miscible CO2 water-alternating-gas injection in the Bakken formation. Energy & Fuels. 2014;**28**(11):6811-6819

[81] Bennion DB, Bachu S. Permeability and relative permeability measurements at reservoir conditions for CO2-water systems in ultra low permeability confining caprocks. In: EUROPEC/EAGE Conference and Exhibition. London, U. K: Society of Petroleum Engineers; 2007

[82] Bennion B, Bachu S. Relative permeability characteristics for supercritical CO2 displacing water in a variety of potential sequestration zones. In: SPE Annual Technical Conference and Exhibition. Dallas, Texas: Society of Petroleum Engineers; 2005

[83] Ramachandran KP, Gyani ON, Sur S. Immiscible hydrocarbon WAG: Laboratory to field. In: SPE Oil and Gas India Conference and Exhibition. Mumbai, India: Society of Petroleum Engineers; 2010

[84] Skauge A, Aarra M. Effect of wettability on the oil recovery by WAG. In: Proceedings 7th European Symposium on Improved Oil Recovery. Moscow: SPE; 1993;**2**:452-459

[85] Skauge A. Summary of Core Flood Results in Connection with WAG Evaluation. in the Tenth Wyoming Enhanced Oil Recovery Symposium. Laramie, Wyoming: University of Wyoming; 1994

[86] Olsen G, Skauge A, Stensen J. Evaluation of the potential application on the WAG process in a North Sea reservoir. In: IOR 1991-6th European Symposium on Improved Oil Recovery. Stavanger, Norway: European Association of Geoscientists & Engineers; 1991

[87] Skauge A and JA. Three-phase relative permeabilities and trapped gas measurements related to WAG processes. in SCA 9421. In: Proceedings of the International Symposium of the Society of Core Analysts. Norway: Stavanger; 1994

[88] Al-Shuraiqi H, Muggeridge A, Grattoni C. Laboratory investigations of first contact miscible WAG displacement: The effects of WAG ratio and flow rate. In: SPE International Improved Oil Recovery Conference in Asia Pacific. Kuala Lumpur: Malaysia Society of Petroleum Engineers; 2003

[89] Srivastava J, Mahli L. Water alternating gas (WAG) injection a novel EOR technique for mature light oil fields a laboratory investigation for GS-5C sand of gandhar field. In: A Paper Presented in Biennial International Conference and Exposition in Petroleum Geophysics; Hyderabad. Pakistan; 2012

[90] Christensen JR, Stenby EH, Skauge A. Review of WAG field experience. In: International Petroleum Conference and Exhibition of Mexico. Villahermosa, Mexico: Society of Petroleum Engineers; 1998

[91] Faisal A, Bisdom K, Zhumabek B, Zadeh AM, Rossen WR. Injectivity and gravity segregation in WAG and SWAG enhanced oil recovery. In: SPE Annual Technical Conference and Exhibition. New Orleans, Louisiana: Society of Petroleum Engineers; 2009

[92] Jenkins MK. An analytical model for water/gas miscible displacements. In:

SPE Enhanced Oil Recovery Symposium. Tulsa, Oklahoma: Society of Petroleum Engineers; 1984

[93] Holloway HD, Fitch RA. Performance of a miscible flood with alternate gas-water displacement. Journal of Petroleum Technology. 1964;**16**(04)

[94] Holm LW. Propane-gas-water miscible floods in watered-out areas of the adena field, Colorado. Journal of Petroleum Technology. 1972; **24**(10)

[95] Cuesta J, Merritt G. Caroline wag project, injectivity and interference test-A field example. In: Journal of Canadian Petroleum Technology. Calgary, Canada: Canadian Inst Mining Metallurgy Petroleum; 1982

[96] Moffitt PD, Zornes DR. Postmortem analysis: Lick creek meakin sand unit immiscible CO2 Waterflood Project. In: SPE Annual Technical Conference and Exhibition. Washington, D.C.: Society of Petroleum Engineers; 1992

[97] Slotte P, Stenmark H, Aurdal T. Snorre WAG pilot. RUTH. 1992;**1995**: 85-91

[98] Surguchev L. WAG Injection. Stavanger: RF-Rogaland Research; 1990. p. 90

[99] Huang ETS, Holm LW. Effect of WAG injection and rock wettability on oil recovery during CO2 flooding. SPE Reservoir Engineering. 1988;**3**(01): 119-129

[100] Stalkup FI. Displacement of oil by solvent at high water saturation. Society of Petroleum Engineers Journal. 1970; **10**(04):337-348

[101] Raimondi P, Torcaso MA. Distribution of the oil phase obtained upon imbibition of water. Society of Petroleum Engineers Journal. 1964; **4**(01):49-55

[102] Bedrikovetsky P. WAG displacements of oil-condensates accounting for hydrocarbon ganglia. Transport in Porous Media. 2003;**52**(2): 229-266

[103] Zahoor M, Derahman M, Yunan M. WAG process design—An updated review. Brazilian Journal of Petroleum and Gas. 2011;**5**(2)

[104] Dorostkar M, Mohebbi A, Sarrafi A, Soltani A. A laboratory study of hot WAG injection into fractured and conventional sand packs. Petroleum Science. 2009;**6**(4):400-404

[105] Selamat SB, Shamsuddin S, Ainatrah A, Halim NA. Evaluation and optimization of enhanced oil recovery by WAG injection at tapis and Guntong Fields, Malaysia. In: SPE Enhanced Oil Recovery Conference. Kuala Lumpur, Malaysia: Society of Petroleum Engineers; 2011

[106] Bond D, Holbrook I, Lack C. Gas drive oil recovery process. US2866507 1958

[107] Li RF, Le Bleu RB, Liu S, Hirasaki GJ, Miller CA. Foam mobility control for surfactant EOR. In: SPE Symposium on Improved Oil Recovery. Tulsa, Oklahoma, USA: Society of Petroleum Engineers; 2008

[108] Xu X. A Novel CO2 flooding based EOR for sandstone reservoirs [doctoral dissertation]. Curtin University; 2016

[109] McLendon WJ, Koronaios P, Enick RM, Biesmans G, Salazar L, Miller A, et al. Assessment of

CO2-soluble non-ionic surfactants for mobility reduction using mobility measurements and CT imaging. Journal of Petroleum Science and Engineering. 2014;**119**:196-209

[110] Schievelbein VH. Method for decreasing mobility of dense carbon dioxide in subterranean formations. U.S. 5033547. 1991

[111] AlSumaiti AM, Hashmet MR, AlAmeri WS, Anto-Darkwah E. Laboratory study of CO2 foam flooding in high temperature, high salinity carbonate reservoirs using co-injection Technique. Energy & Fuels. 2018;**32**(2): 1416-1422

[112] Haruki M, Yawata H, Nishimoto M, Tanto M, Kihara S-I, Takishima S. Study on phase behaviors of supercritical CO2 including surfactant and water. Fluid Phase Equilibria. 2007;**261**(1):92-98

[113] Ryoo W, Webber SE, Johnston KP. Water-in-carbon dioxide microemulsions with methylated branched hydrocarbon surfactants. Industrial & Engineering Chemistry Research. 2003;**42**(25):6348-6358

[114] Fan X, Potluri VK, McLeod MC, Wang Y, Liu J, Enick RM, et al. Oxygenated hydrocarbon ionic surfactants exhibit CO2 solubility. Journal of the American Chemical Society. 2005;**127**(33):11754-11762

[115] Chen Y, Elhag AS, Poon BM, Cui L, Ma K, Liao SY, et al. Ethoxylated cationic surfactants for CO2 EOR in high temperature, high salinity reservoirs. In: SPE Improved Oil Recovery Symposium. Tulsa, Oklahoma, USA: Society of Petroleum Engineers; 2012

[116] Kuehne D, Frazier R, Cantor J, Horn W Jr. Evaluation of surfactants for CO2 mobility control in dolomite reservoirs. In: SPE/DOE Enhanced Oil Recovery Symposium. Tulsa, Oklahoma: Society of Petroleum Engineers; 1992

[117] Radke CJ, Gillis JV. A dual gas tracer technique for determining trapped gas saturation during steady foam flow in porous media. In: SPE Annual Technical Conference and Exhibition. New Orleans, Louisiana: Society of Petroleum Engineers; 1990

[118] Gauglitz A, Friedmann PF, Kam SI, Rossen WR. Foam generation in homogeneous porous media. Chemical Engineering Science. 2002;**57**(19): 4037-4052

[119] Blackwell R, Rayne J, Terry W. Factors influencing the efficiency of miscible displacement. Society of Petroleum Engineers Journal. 1959; **2017**(01):1-8

[120] Xu Q. Theoretical and experimental study of foam for enhanced oil recovery and acid diversion [doctoral dissertation]. University of Texas; 2003

[121] Schramm LL. Surfactants: Fundamentals and Applications in the Petroleum Industry. United Kingdom: Cambridge University Press; 2000

[122] Schramm LL. Foam sensitivity to crude oil in porous media. ACS Advances in Chemistry Series. 1994;**242**:165-200

[123] Gillis J, Radke C. A dual-gas tracer technique for determining trapped gas saturation during steady foam flow in porous media. In: Annual Technical Conference and Exhibition. New Orleans, Louisiana: SPE; 1990

[124] Llama O. Mobility control of chemical EOR fluids using foam in highly fractured reservoirs [MS thesis]. The University of Texas at Austin: Faculty of the Graduate School; 2011

[125] Friedmann F, Jensen J. Some parameters influencing the formation and propagation of foams in porous media. In: SPE California Regional Meeting. Oakland, California: Society of Petroleum Engineers; 1986

[126] Nguyen QP, Alexandrov AV, Zitha PL, Currie PK. Experimental and modeling studies on foam in porous media: A review. In: SPE International Symposium on Formation Damage Control. Lafayette: Louisiana Society of Petroleum Engineers; 2000

[127] Wellington S, Vinegar H. CT studies of surfactant-induced CO2 mobility control. In: SPE Annual Technical Conference and Exhibition. Las Vegas, Nevada: Society of Petroleum Engineers; 1985

[128] Zhang Y, Luo P, Huang S-SS. Improved heavy oil recovery by CO2 injection augmented with chemicals. In: International Oil and Gas Conference and Exhibition in China. Beijing, China: Society of Petroleum Engineers; 2010

[129] Farajzadeh R. Enhanced Transport Phenomena in CO2 Sequestration and CO2 EOR. Netherlands: Delft University of Technology; 2009

[130] Rao D. Gas injection EOR-A new meaning in the new millennium. Journal of Canadian Petroleum Technology. 2001;**40**(02)

[131] Charanjit R, Bernard GG. Method of secondary recovery employing successive foam drives of different ionic characteristics. US3323588A. 1967

[132] Marsden S. Foams in Porous Media; DOE/SF/11564-15; SUPRI-TR-49. Stanford Univ., CA (USA): Petroleum Research Inst.; 1986

[133] Ayirala SC. Surfactant-induced relative permeability modifications for oil recovery enhancement [LSU master's theses]. USA: Louisiana State University; 2002

[134] Zanganeh MN. Simulation and Optimization of Foam EOR Processes. Netherlands: Delft University of Technology; 2011

[135] Khatib Z, Hirasaki G, Falls A. Effects of capillary pressure on coalescence and phase mobilities in foams flowing through porous media. SPE Reservoir Engineering. 1988;**3**(03): 919-926

[136] Zanganeh MN, Kam SI, LaForce T, Rossen WR. The method of characteristics applied to oil displacement by foam. SPE Journal. 2011;**16**(01):8-23

[137] Adkins SS, Chen X, Chan I, Torino E, Nguyen QP, Sanders AW, et al. Morphology and stability of CO2-in-water foams with nonionic hydrocarbon surfactants. Langmuir. 2010;**26**(8):5335-5348

[138] Liu Y, Grigg RB, Bai B. Salinity, pH, and surfactant concentration effects on CO2-foam. In: SPE International Symposium on Oilfield Chemistry. The Woodlands, Texas: Society of Petroleum Engineers; 2005

[139] Handy LL, Amaefule JO, Ziegler VM, Ershaghi I. Thermal stability of surfactants for reservoir application (includes associated papers 11323 and 11597). Society of Petroleum Engineers Journal. 1982;**22**(05)

[140] Mayberry DJ, Kam SI. The use of fractional-flow theory for foam displacement in presence of oil. SPE Reservoir Evaluation & Engineering. 2008;**11**(04):707-718

[141] Nikolov A, Wasan D, Huang D, Edwards D. The effect of oil on foam stability: Mechanisms and implications for oil displacement by foam in porous media. In: SPE Annual Technical Conference and Exhibition. New Orleans: Louisiana Society of Petroleum Engineers; 1986

[142] Law D-S, Yang Z-M, Stone T. Effect of the presence of oil on foam performance: A field simulation study. SPE Reservoir Engineering. 1992;7(02): 228-236

[143] Chou SI, Vasicek SL, Pisio DL, Jasek DE, Goodgame JA. CO2 foam field trial at North Ward-Estes. In: SPE Annual Technical Conference and Exhibition. Washington, D.C.: Society of Petroleum Engineers; 1992

[144] Enick RM, Olsen DK. Mobility and Conformance Control for Carbon Dioxide Enhanced Oil Recovery (CO2-EOR) via Thickeners, Foams, and Gels—A Detailed Literature Review of 40 Years of Research; DOE/NETL-2012/1540; U.S. Tulsa: Dept. of Energy, National Energy Technology Laboratory (NETL); 2011

[145] Jonas TM, Chou SI, Vasicek SL. Evaluation of a CO2 Foam field trial: Rangely weber sand unit. In: SPE Annual Technical Conference and Exhibition. New Orleans, Louisiana: Society of Petroleum Engineers; 1990

[146] Stephenson DJ, Graham AG, Luhning RW. Mobility control experience in the Joffre viking miscible CO2 flood. SPE Reservoir Engineering. 1993;**8**(03)

[147] Heller J, Boone D, Watts R. Testing CO2-foam for mobility control at rock creek. In: SPE Eastern Regional Meeting. Morgantown: West Virginia Society of Petroleum Engineers; 1985

[148] Stevens J. CO2 foam field verification pilot test at EVGSAU: Phase IIIB—Project operations and performance review. SPE Reservoir Engineering. 1995;**10**(04):266-272

[149] Mukherjee J, Nguyen QP, Scherlin J, Vanderwal P, Rozowski P. CO2 foam pilot in salt creek field, Natrona County, WY: Phase III: Analysis of pilot performance. In: SPE Improved Oil Recovery Conference. Tulsa, Oklahoma, USA: Society of Petroleum Engineers; 2016

[150] Sanders A, Jones RM, Rabie A, Putra E, Linroth MA, Nguyen QP. Implementation of a CO2 foam pilot study in the SACROC field: Performance evaluation. In: SPE Annual Technical Conference and Exhibition. San Antonio, Texas, USA: Society of Petroleum Engineers; 2012

[151] Janssen A. Enhanced fluid separation for chemical floods. In: Proceedings of the SPE Chemical Flooding-EOR Workshop; Penang, Malaysia; May 2012

[152] Razali H. General requirement of chemical EOR from HSE perspectives. In: Proceedings of the SPE-Chemical Flooding-EOR Workshop; May 2012; Penang, Malaysia

[153] Waldman T. Design considerations for chemical EOR facilities. In: Proceedings of the SPE-Advanced Technologies Workshop; May. 2012; Penang, Malaysia

[154] Zain T. Storage, logistic and chemical transferring in offshore environment. In: Proceedings of the SPE-Chemical Flooding-EOR Workshop; May. 2012; Penang, Malaysia

[155] Henderson J, Meyer W, Taber J. Miscible drive secondary oil recovery process. U.S. Patent 3,330,345. 1967

[156] Jianhang X. Carbon Dioxide Thickening Agents for Reduced CO2 Mobility. Pittsburgh: University of Pittsburgh; 2003

[157] Lee J, Cummings S, Dhuwe A, Enick R, Beckman E, Perry R, et al. Development of small molecule CO2 thickeners for EOR and fracturing. In: SPE Improved Oil Recovery Symposium. Tulsa, Oklahoma, USA: Society of Petroleum Engineers; 2014

[158] Kirby CF, McHugh MA. Phase behavior of polymers in supercritical fluid solvents. Chemical Reviews. 1999; **99**(2):565-602

[159] Enick RM, Beckman EJ, Shi C, Huang Z, Xu J, Kilic S. Direct thickeners for carbon dioxide. In: SPE/DOE Improved Oil Recovery Symposium. Tulsa, Oklahoma: Society of Petroleum Engineers; 2000

[160] Enick RM, Beckman EJ, Hamilton A. Novel CO2-Thickeners for Improved Mobility Control. U.S. Department of Energy Office of Scientific and Technical Information: University of Pittsburgh (US); 2001

[161] Wang Y, Hong L, Tapriyal D, Kim IC, Paik I-H, Crosthwaite JM, et al. Design and evaluation of nonfluorous CO2-soluble oligomers and polymers. The Journal of Physical Chemistry B. 2009;**113**(45): 14971-14980

[162] Tapriyal D. Design of non-fluorous carbon dioxide soluble compounds [doctoral dissertation]. University of Pittsburgh; 2009

[163] Zhang J, Xiao B, Zhang G, He W, Yang G, Jing T, et al. A new thickener for CO2 anhydrous fracturing fluid. In: MATECP Web of Conferences. Romania: EDP Sciences; 2015

[164] Dhuwe A. Thickeners for Natural Gas Liquids to Improve the Performance in Enhanced Oil Recovery and Dry Hydraulic Fracking. USA: University of Pittsburgh; 2016

[165] Kilic S, Michalik S, Wang Y, Johnson JK, Enick RM, Beckman EJ. Phase behavior of oxygen-containing polymers in CO2. Macromolecules. 2007; **40**(4):1332-1341

[166] Shi C, Huang Z, Beckman EJ, Enick RM, Kim S-Y, Curran DP. Semi-fluorinated trialkyltin fluorides and fluorinated telechelic ionomers as viscosity-enhancing agents for carbon dioxide. Industrial & Engineering Chemistry Research. 2001;**40**(3): 908-913

[167] O'Brien MJ, Perry RJ, Doherty MD, Lee JJ, Dhuwe A, Beckman EJ, et al. Anthraquinone siloxanes as thickening agents for supercritical CO2. Energy & Fuels. 2016;**30**(7):5990-5998

Chapter 4

Minimum Miscibility Pressure, Miscible Displacement, CO_2 Capture, and Injection

Julio Gonzalo A. Herbas Pizarro

Abstract

The Minimum Miscibility Displacement Pressure, and the strategies to maintain the reservoir pressure above the minimum miscibility pressure are the most important elements for a successful EOR Dry Gas, CO_2, or N_2 Miscible injection project. The Miscibility behaviour needs to be understood early after the reservoir discovery to establish if a miscible displacement is economically attractive. The difference of a miscible gas displacement with an immiscible displacement is of such importance because a miscible displacement could achieve a recovery factor as high as 75% to 90% of the contacted oil compared to 30–40% recovery factor for an immiscible displacement process. In some field cases, the MMP is determined in the mid or late field life when the reservoir pressure, temperature and fluids distribution might limit the time left to design and implement a miscible gas displacement; in other, the operators possess the technology to design and implement Miscible Gas Displacement and the ability to articulate the project economy allowing time on decisions to implement, operate, and materialize the incremental recovery from a miscible displacement; therefore, it is recommended to determine the miscibility pressure, as soon the field is identified as candidates for EOR.

Keywords: EOR, CO_2, miscible displacement, CCS

1. Introduction

This chapter discuss the concepts and elements that drive a Miscible Displacement, some practical strategies for project design, implementation and evaluation, field experiences from the Minimum Miscibility Pressure concept, application, and influence in field cases of Miscible Gas Injection projects performance, including natural gas, Dioxide Carbonate (CO_2), Nitrogen (N_2) and Flue. Some field cases of EOR Miscible Displacements injecting dry gas and CO_2 are also discussed in the context of EOR operations.

Historically it is more common to deal with immiscible gas injection projects, compared with the cases of miscible gas injection projects possibly because the opportunities for implementation of miscible displacement have not been identified in early stages, the high costs of compression to achieve miscibility and the access to the know-how.

There have been cases where the implemented reservoir management strategy was focused to let the reservoir pressure to deplete below the bubble gas pressure to create a secondary gas cap to use the gas cap expansion as production mechanism. This strategy might be considered reasonable and economic; however the cases that we have seen have recoveries in the range of 35–45% of the original oil in place at the time when the GOR has increased to extremely high values that suggest the injected gas is being recycled. The recovery factor in those cases might had been in order of 60–70% if a miscible gas injection process would have been implemented at early stage of the field life cycle.

The determination of the MMP can be estimated with reasonable accuracy if there is available a compositional analysis of the reservoir oil and a representative PVT analysis, which can be used to build a representative one-dimension compositional simulations for various types of gases that might be available for injection. Usually it is important a survey of potential gas sources in the area. Once the MMP has been estimated by compositional reservoir simulation, the next step is to verify the model work with laboratory experiments applying methods such as slim tube tests, rising bubble, zero interfacial tension, these last two are more recent developments in determination of MMP.

The typical candidate gases for injection are dry or wet natural gases, Nitrogen, CO_2 and flue gas, a product from natural gas combustion; from those gases, the CO_2 has been identified as the more efficient miscible agent based in its property to dissolve the oil.

The current trend of Carbon Capture and Storage (CCS) objectives pursued by the industry to reduce the green house effects can be levered with the implementation of more CO_2 miscible injection projects elsewhere the CO_2 is available, as there are several oil fields that have not developed because the hight CO_2 content.

An early evaluation of the economic feasibility to achieve a miscible displacement is of paramount importance which should be followed with the formulation of a doable strategic implementation plan for the project construction to materialize the incremental recovery factor and the incremental production, that in turn is dependent of a sound reservoir management conscious of the project objectives, that works with open communication between and with participation of all the company players from top management to field engineers and operators.

2. Generalities of gas injection miscible displacement process

The crestal gas injection is one of the more efficient traditional displacements processes, it works by the gravity segregation drive mechanism displacing the oil downward toward the producer wells placed down in the structure. A miscible displacement is if not the most efficient, one of the more efficient displacement processes because the injectant fluid dissolves the oil as the injectant at displacement front gets in contact with the oil in the reservoir zones, once it gets in contact; therefore, the displacement front not only displaces the movable oil saturation but also dissolves the residual oil saturation that is typically left behind in an immiscible gas injection process. Therefore, the oil saturation behind the displacement front in a Miscible Displacement can be as very low virtually zero.

3. What is miscibility

Miscibility is the mixture of two fluids, one fluid dissolves a second fluid either at first contact or in multiple consecutive stages as the injectant fluid contact and displace the second fluid. First, contact miscibility is driven by the fluid s composition and the thermodynamic conditions: pressure and temperature.

An important feature of the Miscible Gas injection Displacement is the solubility effect of the gas displacing miscible oil which eliminates the gas oil relative permeabilities effect, which is a consequence of the dissolution of the displacing fluid into de displaced fluid, that convert the displacement as one uniform front moving at the interface of gas displacing oil.

The **Figure 1** shows gas (CO_2) miscible displacement of trapped oil in a porous media, the CO_2 gas mix with the oil, swells the oil molecules, and extract light components from the oil as it moves into the reservoir, creating a virtual wash of the porous media.

Other characteristics of a miscible gas displacement:

- A miscible gas displacement can remove and displace the trapped oil in the porous, which is not movable by immiscible displacement.
- Miscible displacement can be achieved injecting lean gas, wet gas, (C1, C1–C2, CO_2, etc.).
- At minimum miscibility pressure (MMP), the interfacial tension between the oil and the displacing fluid is approaching zero.

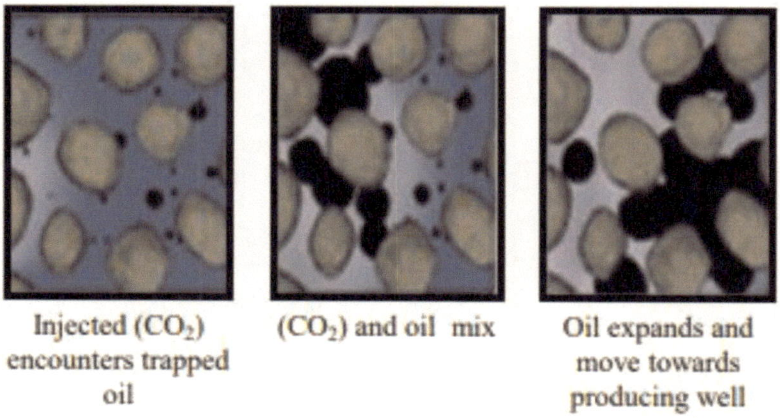

Figure 1.
CO_2 injection miscible displacement in pore scale [1].

- Under normal conditions, oil & gas reservoir fluids form distinct, immiscible phases; the immiscible phases are separated by an interface associated with interfacial tension (IFT), when the interfacial tension is equal to zero the two fluids mix achieving the miscible condition.
- The residual oil saturation to gas (and water) is directly proportional to IFT
- A miscible displacement is characterized by low/zero residual oil saturations.

Miscible displacement processes can be implemented in absence of structural dip as it is shown in **Figure 2**. In dipping reservoirs, the gravity segregation will favour a stable displacement as described in the Field case. Early Identification of Multiple Contact Miscibility injecting Dry Gas El Furrial Field, a Case of Multiple Contact Miscible Gas Injection combined with Low-Salt Water Injection.

4. Types of miscible process and mechanisms

There are two basic types of miscibility:

4.1 First contact miscibility

Occurs when the injectant dissolves the oil as soon it gets in contact with the reservoir, it usually occurs with solvents as gasolines, and very rich gases.

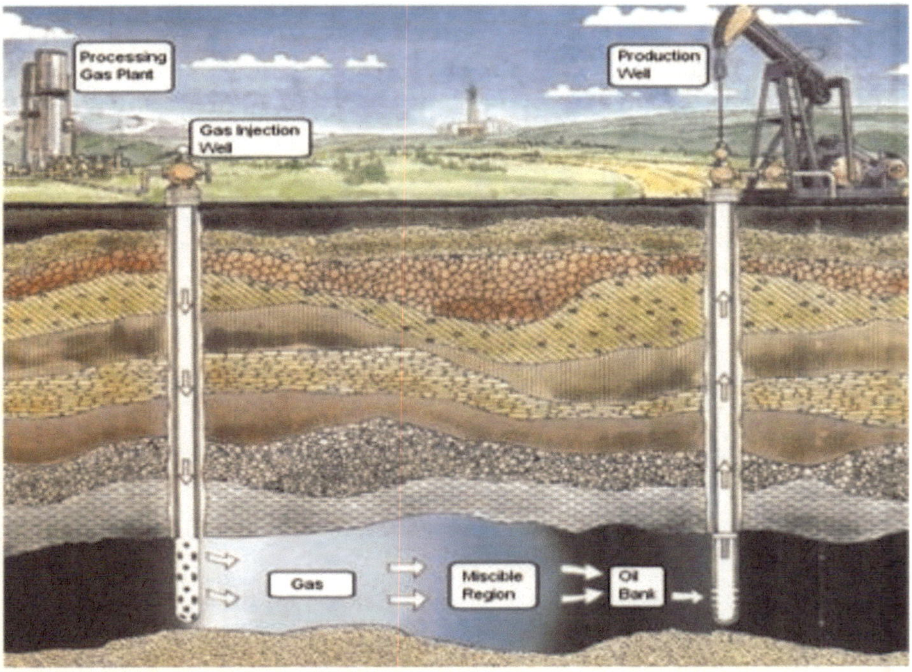

First Contact Miscibility

Figure 2.
Miscible gas injection in absence of structural dip.

4.2 Multiple contact miscibility (MCM)

A multi contact miscibility starts as an immiscible displacement, then the thermodynamic conditions (Pressure and Temperature) allow a continuous transfer of molecules of hydrocarbon from the displaced oil to the injectant (displacing phase), in a condensing and vaporising process, that enrich continuously the injected gas, until it becomes miscible with the displaced oil. The mechanic of miscibility injecting a dry gas is defined as Vaporizing Drive, it is controlled by the oil composition, the pressure, and the temperature.

The miscibility achieved through multiple contacts between the injection gas and the oil in-place occurs after the injection gas at the displacement front progressively contact the oil in the reservoir. As the displacement front moves into the reservoir the gas takes more heavier components until the miscibility is achieved. In presence of viscous fingering or permeability heterogeneity, the minimum distance to accomplish miscibility increases because of dispersion at the displacement front. The total recovery in a MCM process is the sum of the recovery obtained while injection gas travels the immiscible portion of the porous media plus the recovery obtained when the gas displaces the miscible portion.

In a phase envelope **Figure 3** the first contact miscibility pressure usually occurs above the bubble point pressure. The Cricondenbar is the maximum pressure that gas phase cannot be formed any more regardless of its temperature, its temperature is called cricondenbar temperature.

The Cricondentherm is the maximum temperature that liquid cannot be formed regardless of pressure and its pressure is called cricondentherm pressure.

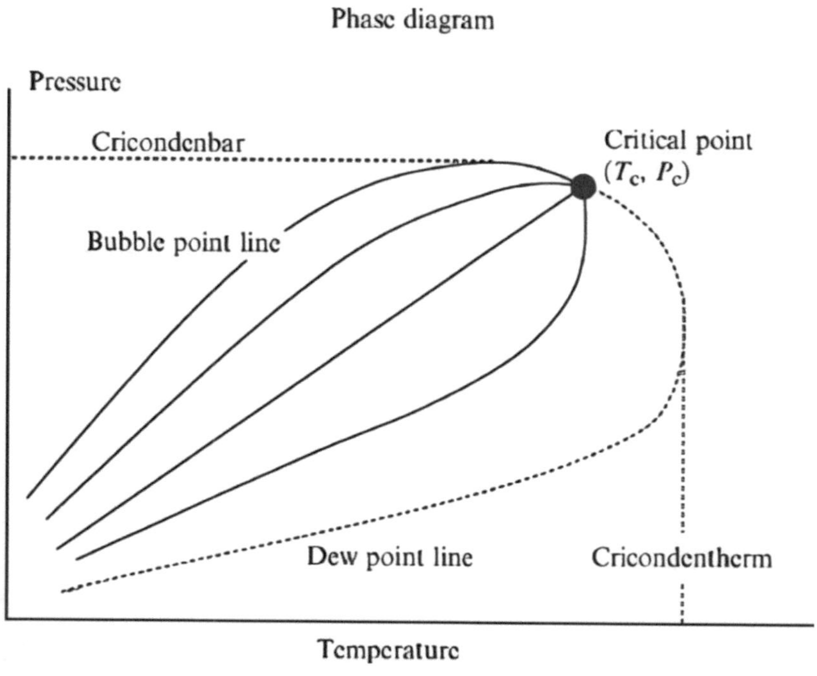

Figure 3.
Pressure volume phase diagram for a typical oil composition [2].

At temperatures higher than Cricondentherm, only one phase occurs at any pressure, the corresponding pressure is called Cricondenbar that is the maximum pressure above which no gas can be formed regardless of the temperature.

The chart Pressure Temperature phase diagram for an oil system illustrates the position of the cricondenbar pressure, the bubble point line and the two phases liquid and gas envelop; the area above the bubble point line and below the cricondenbar pressure defines the region where the multiple contact miscibility might take place.

5. Ternary diagrams

Ternary diagrams are used to represent the phase behaviour of hydrocarbon systems, the mixture of components of the fluids with more than three components is divided into three pseudo components, such as light, intermediate, and heavy components of a hydrocarbon phase. The ternary diagrams are developed based on compositional equations of state (EOS) developed to replicate the phase behaviour of hydrocarbon systems, are useful to represent the phase behaviour of a mixture of pure hydrocarbons. The composition of the 3 points in ternary diagrams is defined based on the oil composition and the injection fluid composition., and the diagrams are generated with specialized commercial software as Eclipse 300, GEM, etc. The grouping of components is usually a convention defined by the user

- Thermodynamic criteria define the minimum miscibility pressure MMP in a ternary diagram as the pressure at which the limiting tie line passes through the point representing the oil composition.

- The tie lines in a ternary diagram, represents the two fluids being mixed, compositional concentrations of the two mixed fluids are given by the ends of the tie lines.

- The mixture composition lies on the line, its position depends on the concentration ratio of the two end-point fluids.

As for example in **Figure 4**, a gas composition 50% C1 and 50% C2–C4 will lie in the midpoint of the C1 and C2–4 corners.

The composition of an Oil sample in the centre of the triangle represents a mixture of:

- 30% C1 with 70% C2–4

- 30% C7+7 with 30%C2–4

- 37% C7+ with 3% C1

The green area represents the mixture of 3 groups C1, C2–4 and C7+ result in 2 phases gas and liquid.

The Triangular Diagrams in **Figure 5** represents a system with C1 at the top corner, C2–C4 at the bottom right corner and C7+ at the left corner. A point between two corners represents a composition proportional to the corners, as example a midpoint between C1 and C4 represent a mixture 50% C1 and 50% C2–4; similarly the midpoint

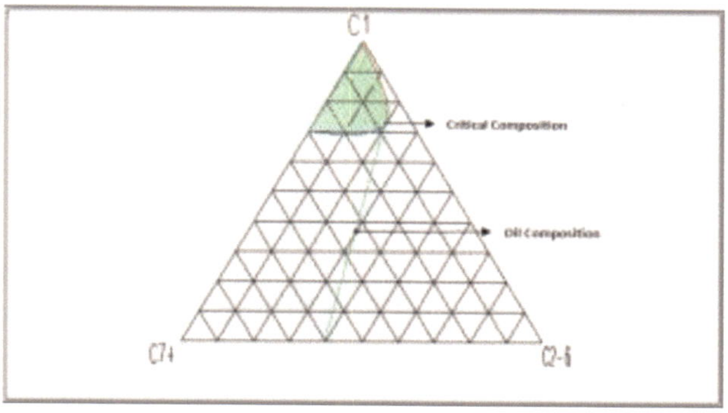

Figure 4.
Typical Ternary phase diagram, hydrocarbon system: the limiting tie line passes through the oil composition at minimum miscibility pressure (MMP) [1].

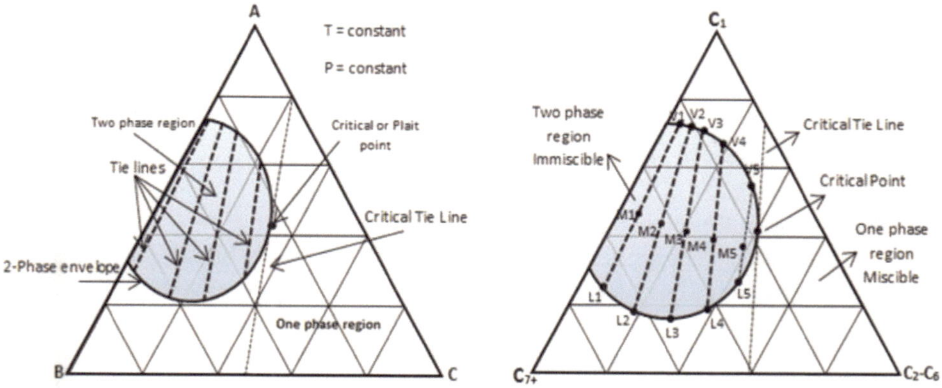

(a). ternary phase diagram for a system of components A, B, C with limited miscibility

(b). Pseudoternary diagram

Figure 5.
Ternary diagrams system with limited miscibility [3].

between C4–C2 and C7+ represent 50% of each corner. For this composition, the blue areas correspond to the mixtures that result in 2 phases oil and gas. The miscibility occurs when a gas composition gets in contact with an oil composition without crossing the two phases area.

A multiple contact miscible process MMP with dry gas is illustrated in 4 steps in the phase diagram in **Figure 6**.

1. The injection gas is composed by pure C1, it is represented with the point "G" in the diagram. As the C1 gas moves in the reservoir, thermodynamic interactions with the oil components, makes the C1 extract intermediate components such as C2–C4 etc. from the contacted oil phase.

2. As the gas move forward, the intermediate molecular weight hydrocarbons from the injected gas are transferred into the oil in the reservoir.

3. Them, because of the phase transfer mechanism some of the gas "condenses" into the oil.

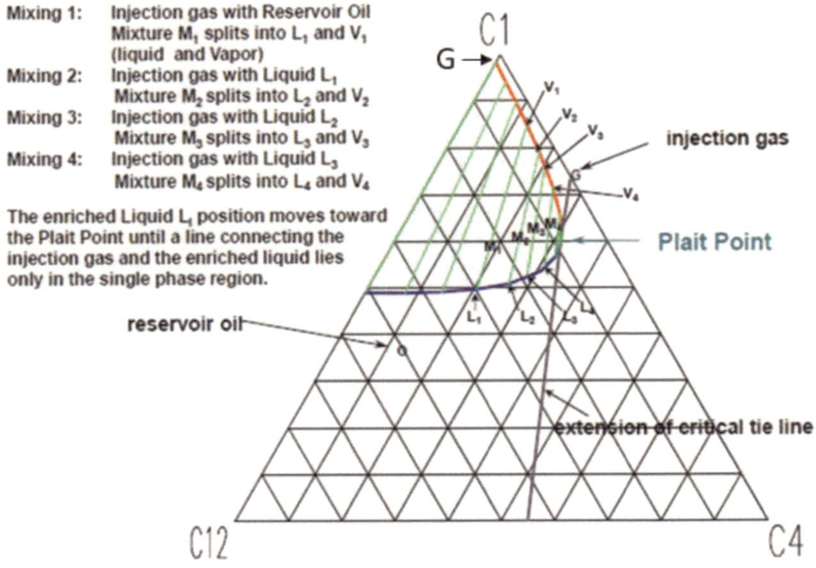

Figure 6.
Three phases' diagrams condensing gas drive miscibility [3].

4. The reservoir oil becomes enriched with these materials, until the miscibility occurs between the injection gas that has already extracted high Molecular weight components from the residual oil and the enriched oil at the displacement front.

The plait point is a critical point at which the liquid and vapor phases are identical.

The miscibility occurrence is a function of the solvent concentration, (C4). The injectant composed by pure C4 and high C4 concentration achieve first contact miscibility. As the proportion of the solvent C–4 reduces, there will be a composition where there is not first contact miscibility, and miscibility might occurs by multiple contact, then as the C4 proportion reduces the miscibility is lower until a point where there is no miscibility at all, as shown in **Figure 7**.

6. Condensing—Gas mechanism sequence

1. Reservoir pressure is lower than the MMP (minimum miscibility pressure)

2. The solvent (injectant) and the oil are not miscible initially.

3. As the dry gas move in the reservoir there is transfer of hydrocarbon components from the residual oil to the injectant and from the solvent components transfer to liquid oil phase.

4. Repeated contact between oil and solvent moves system towards the plait (critical) point (dynamic miscibility).

As the pressure increases the two-phase region becomes smaller. At some pressure the injected gas is to the right of the limiting tie line and MCM develops. This process is known as condensing vaporizing multiple contact miscible drive.

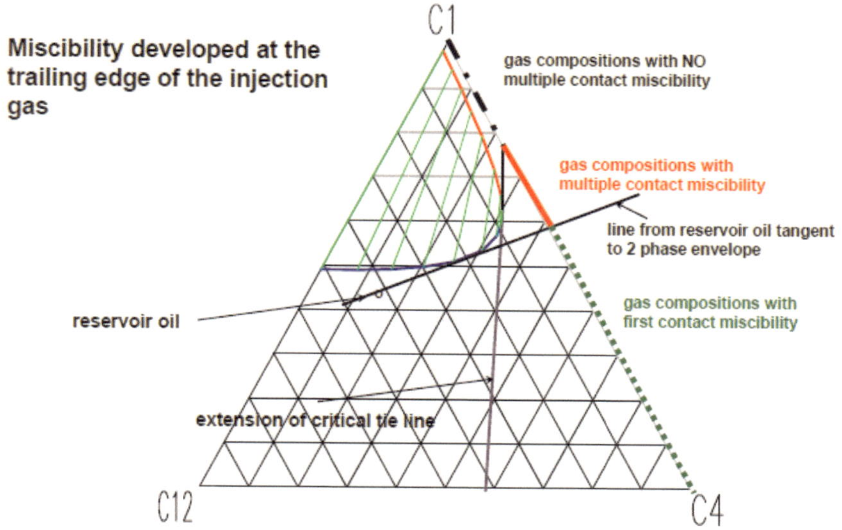

Figure 7.
Three phases diagrams showing gas composition with multiple MCM and gas composition with first contact miscibility FCM [4].

7. Vaporizing gas mechanism

Intermediate hydrocarbon components in the oil vaporize to enrich the gas. As the leading edge of the gas slug becomes sufficiently enriched, it becomes miscible with the reservoir oil.

7.1 What happens in the simulation when miscibility is achieved?

When the miscibility is achieved into the reservoir, the displacement will be very efficient and virtually all the oil in the reservoir will be removed and displaced toward the producer wells, the characteristic of miscible displacement projects is the high recovery factors in order of 60–75% of the oil in place; higher figures of the recovery factor are not commonly reported, limited by the operational pressure in the field, or lack of gas injection continuity, reservoir heterogeneities, etc.

Experiments show that final recovery increases by increasing the slim tube length for any injection rate.

8. Challenges in miscible gas injection projects

One difficulty in a Miscible Gas Displacement project is to keep all the reservoir porous media above the minimum displacement pressure. Several pressures levels can coexist in the reservoir because of the pressure gradient and the flow dynamics in a heterogeneous reservoir as sedimentary environments composed typically by river channels, plains, sandstones, bars, splays, etc. or in carbonate reservoirs with several facies within the reservoir unit; if there is not a well-defined safety margin above the MMP, there might be areas with pressures below the minimum miscibility pressure.

It is common to receive management requirements to produce at maximum potential which might not correspond to the injection rates designed to maintain the

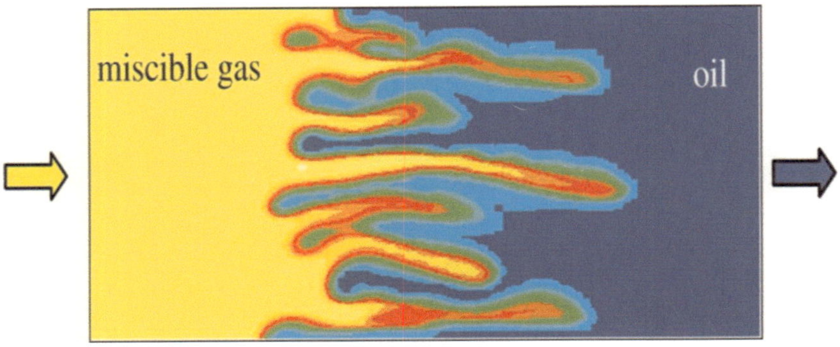

Figure 8.
Fingering in a miscible displacement [2].

reservoir pressure above the MMP, this is a challenge in a MMP project that might cause detrimental effects to the recovery factor.

In the last decades of past century and early times of this century, the oil and gas operators realized the importance of understanding the fluid behaviour and its characterization with application of fluid phase envelops and its use as a reservoir management tool. In many cases, the feasibility to increase dramatically the recovery factors was recognized after understanding the reservoir the reservoir dynamic, in some cases a late implementation of a MMP EOR project was hampered by the high cost of repressurizing the reservoir at levels adequate for a MMP displacement; this type of issues has avoided a more extended implementation of EOR Miscible gas injection projects.

Gravity stable injection of gases into high relief oil reservoirs can result in substantial incremental oil recovery, depending on the densities of the gases at reservoir conditions, the gases should be injected at the crest or bottom of the reservoir, while miscible displacement scan be in low relief of flat reservoirs, however the process will be more challenging because no gravity effects.

Viscous fingering is another challenge that can result in poor vertical and horizontal sweep efficiency (**Figure** 8).

Other challenges are the potential corrosion, affecting the well and the production facilities as also nonhydrocarbon gases must be separated from saleable gas.

9. Typical gases that can be used as injectants

The more common gases for injection are the associated gas produced with the oil, dry gases, or gases available in the nearby of the field's candidate for miscible injection, and other gases as CO_2, N_2, and Flue gas. Flue gas is a mixture of air with combustion gases, with the advantage that the volume of gas used in the combustion is multiplied by several orders of magnitude. Cleaning requirement of impurities in the flue gas will be depending in each case. The miscibility displacement with flue gas usually requires much higher reservoir pressures sometimes to impractical levels.

Usually, it is important to execute a source of injectant fluid study covering for example 100 kilometres around the project location with the purpose to investigate the potential sources of gases for injection. It should be done in a short period by personal of the operators familiar with this type of process. From the gas's availability study, it should be generated the different types of gases to be used in the determination of the optimum injectant for a particular EOR miscible gas injection project.

10. Screening parameters for a miscible project

The typical parameters for a favourable miscible process are shown below, it has been defined in several known publications, as "Updated EOR screening, JJ Taber, F.D. Martin, SPE, and R.S. Seright, SPE, New Mexico Petroleum Recovery Research Center SPE 1997 [5]' and Aladasani, Ahmad, "Updated EOR screening criteria and modeling the impacts of water salinity changes on oil recovery" (2012). Doctoral Dissertations [6].

However, every field case should be studied individually considering all factors inherent to the field as reservoir size, reserves, available fluids for injection, markets, among others.

- Gravity >24°API (35°API for N_2)
- Viscosity <10cp
- Composition C1–C7
- Oil saturation 30%
- Formation type sandstone/carbonate
- Net thickness (thin unless dipping)
- Average formation permeability (not critical)
- Reservoir Transmissibility (not critical)
- Depth >4000ft
- Temperature not critical

The average reservoir permeability is the arithmetic or geometric, weighted average of the permeabilities defined from electrical logs and cores used to populate a reservoir grid. Transmissibility is a term to express the reservoir ability to move fluids as function of relative permeabilities, fluid viscosity, formation volume factor, and geometric parameters.

11. Experimental determination of the minimum miscibility pressure (MMP)

There are several experimental methods to determine the MPP, the more known are the Slim Tube Tests, the Raising Bubble, the Zero Interfacial Tension; en addition there are other methods to estimate the MPP as compositional reservoir simulations and correlations.

Before initiation of the experimental laboratory work, it is recommendable to preliminary estimate the MMP by means of compositional reservoir simulations, this can be done building a simple one-dimensional reservoir simulation model based in the compositional fluid characterization, representative rock properties pressures and temperatures, then displace the oil using various injectants and various reservoir pressures; the obtained results can be used to evaluate the process under several pressures and to determine the conditions required to achieve a miscible displacement.

The one-dimension compositional modelling is recommended to be the first task to be done as soon as the compositional description and the PVT data is available, these will guide the laboratory experiments, and the preparation of full field compositional reservoir simulations studies.

12. Slim tube tests

The slim tube test is an apparatus laboratory test used to estimate the minimum miscibility pressure (MMP) or minimum miscibility concentration (MMC) of a given injection solvent and reservoir oil. It allows to create a porous media saturated with the reservoir fluid at representative pressure and temperature to establish the original conditions prevalent in the reservoir, which will be used to test the injectant as dry gas, wet gas, separator gases, CO_2, Nitrogen, or Flue Gas.

The slim tube **Figure 9** is a narrow long-coiled stainless-steel tube. It is filled with sand of a specific mesh size like the reservoir porous media. To model the displacement process in the reservoir, the tube is saturated with reservoir oil at a reservoir temperature, then the Gas injection is performed at several test predefined pressures, or at the field reservoir pressure if this is a undersaturated reservoir and the

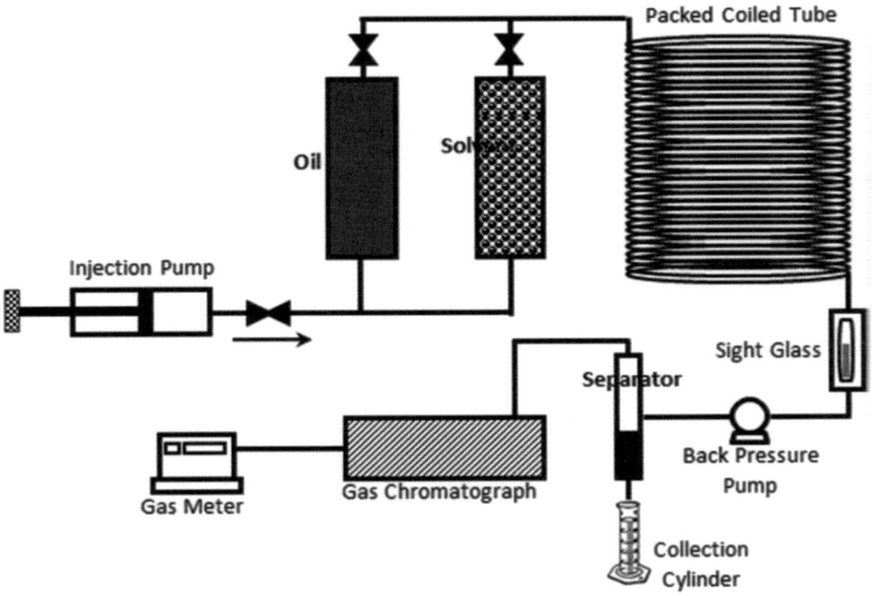

Figure 9.
Slim tube test apparatus schematic [3].

investigation is done aiming to investigate the process at that operational pressure. The produced effluents, density and composition are measured as functions of the injected volume.

A slim tube internal diameter (ID) is typically about 5/16 inches, with length from 5 to 40 meters. The tube is filled up with glass beads or sand of a specific mesh size, the ratio of particle size to tubing diameter is sufficiently small, less than 1/10, to neglect wall effects, it can be idealized as a one-dimensional element of the reservoir.

When gas is injected in the sand packed slim tube apparatus will take place multiple equilibrium contacts, at the end of the experiment the recovery factor is calculated to identify the type of displacement, a miscible displacement will be concluded at recovery factor close to 95% or more. The slim-tube tests result should not be indicative of ultimate recovery to be achieved in actual reservoir, because the Slim Tube is not including factors sweep efficiency, transition zone length, etc. The experiment should be done at constant reservoir temperature.

The experimental procedure requires an initial calibrating of the apparatus with known fully miscible fluids. To determine the MMP, the slim tube is saturated with crude oil and several consecutive displacements are executed at various reservoir pressures.

The oil recovery after injection of a specific number of pore volumes (PV) such as 1.2 PV of solvent is the test criterion for miscibility. The recovery factors for the different pressures are plotted versus pore pressure for the several slim-tube tests, typically at low pressures recoveries will be low, and will increase as the pore pressure increases, when the slop of the first line exhibit a noticeable change, it is indicative of the multiple contact miscible pressure, further higher pressures should reach higher recoveries, **Figure 10**.

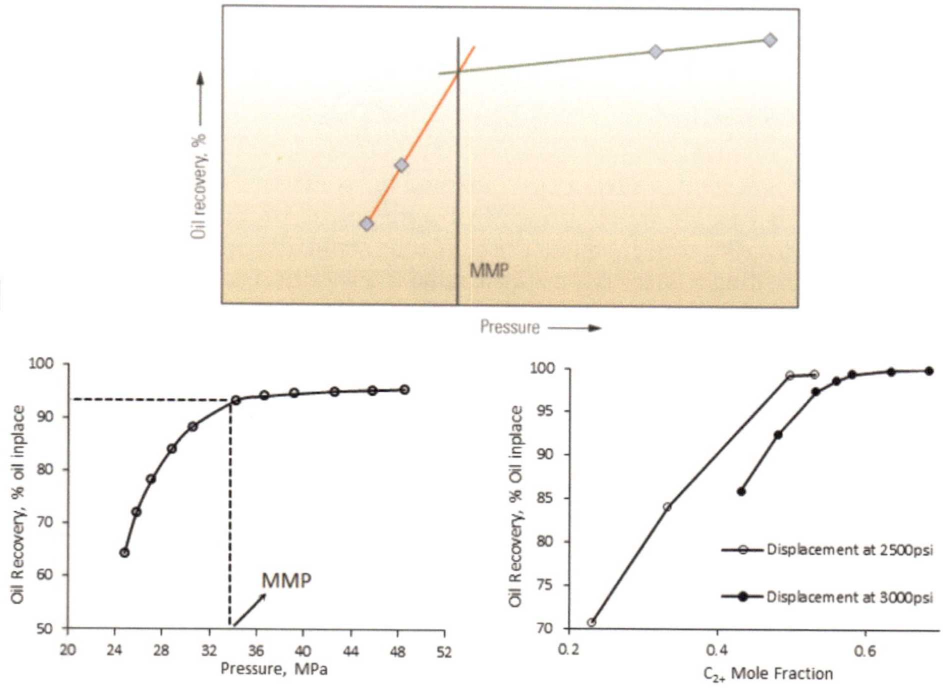

Figure 10.
Plotting results from the slim tube test experiments.

The slim tube test usually considers consecutive displacements at different pressures, starting from the estimated from the MMP compositional simulation, the obtained results for the several pressures are plotted and the tests are repeated until reaching a near 95% recovery factor. The 2 lower plots in **Figure 10** show the MMP determination injecting soltrol (an isoparaffinic solvent) in slim tube saturated with synthetic oil. The experiments can be done with different solvent concentrations to evaluate the solvent minimum requirements.

Different strategies for the determination of the MMP can be designed, as for example, reducing the number displacements to at least four pore pressures, if the results show two trends, the point of intersection of the trends is considered the estimated MMP for the given oil-solvent system. In other cases, a particular reservoir pressure might have been defined as the target operation reservoir pressure to operate the reservoir with specific purpose as to avoid crossing the asphaltene flocculation onset, in that case it is a practice to run the slim tube tests a that specific operational pressure with the objective to understand the process, a particular case where this strategy was successfully applied is described in the field experiences section.

In any strategy that is used, the results obtained from a slim tube test must be used as input to fine-tune an equation of state for reservoir simulation, that will be applied in the full field compositional simulation required to estimate the field recovery factor; the accuracy of the predictions is function of the data representativeness.

The displacement from the reservoir is affected by various mechanisms that causing dispersion, such as gravity override and viscous fingering caused by unfavourable viscosity ratio. The porosity heterogeneity if present will also cause dispersion of the front. The slim tube provides a one-dimensional dispersion free displacement of oil; therefore, the dispersion effects must be studied with a three-dimension multi geocellular model representative of the reservoir. At field condition, the MMP and the final recovery are function of the thermodynamic behaviour in the system, the gravity effects, reservoir heterogeneities, etc.

12.1 Micro slim tube test

Interface Fluidics has created a novel method to estimate miscibility using a microfluidic chip capable of measuring MMP with greater precision and confidence than the standard slim tube method. The slim-tube standard apparatus estimates MMP by constructing a linear regression around a few critical data points, Micro Slim Tube use Interface's analogue a data-driven approach to yield results with greater

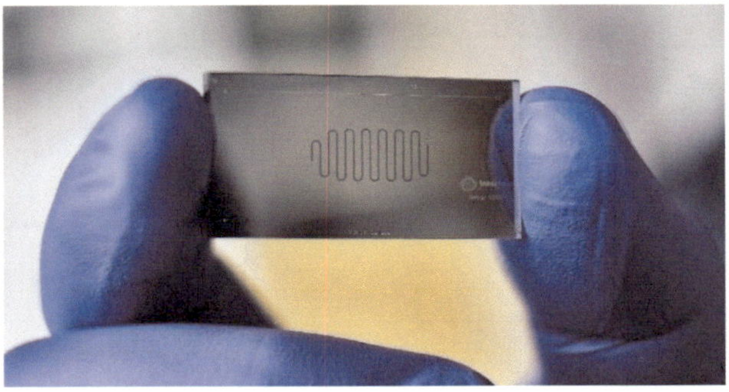

Figure 11.
Micro slim testing by interface fluidics oil.

accuracy. Each chip run collects anywhere from 25 to 75 values per gas-oil system, resulting in high-resolution data plots that can be used to determine MMP directly.

The Micro Slim Tube **Figure 11**, is a miniature of the slim-tube method via microfluidic technology, for rapid and cost-efficient determination of MMP using smaller sample volumes to conduct initial tests. This allows to analyse several reservoirs samples and conditions for the MMP investigation. Changes in oil and gas composition can dramatically impact MMP values, accurately capturing this variable allows to cover a wider spectrum of condition for reservoir simulation models and detailed planning of miscible gas flooding processes.

It can reduce cost for miscibility measurements, and execution of minimum miscibility enrichment (MME) studies, to optimize the gas injection strategy. Requirement of lower sample volumes impact favourably the economy and health/environmental risks associated with sampling from wells.

12.2 Raising bubble method

This s a more recent development laboratory method to indicate miscibility between the reservoir oil and injection gas at specific conditions of pressure and temperature.

- A gas bubble is injected into an oil-filled visual cell at a given temperature and test pressure.

- The progressive change in shape of the rising bubble indicates its miscibility with the oil at those conditions.

- Below the minimum miscibility pressure (MMP), the bubble holds its shape as it rises, above the MMP, the bubble shape changes as it rises.

- It may disintegrate, dissolve, or disappear into the oil.

Testing at several pressures helps to determine the MMP between the gas and oil. The rising-bubble test represents a forward-contacting miscibility process and therefore may not accurately to estimate the MMP for a backward or combined contact mechanism [7].

12.3 Rising bubble apparatus by core laboratories

The Rising bubble apparatus (RBA) offered by Core laboratories provides a fast, accurate, cost-effective measurement of minimum miscibility pressure. The essential feature of the apparatus is a flat glass tube mounted vertically in a high-pressure sight gauge in a temperature-controlled oven.

The glass tube, its **Figure 12** approximately 20 cm long, facilitates the examination of bubbles rising in opaque oils. The glass tube is back lit for visual observation of the tube contents. A hollow needle is mounted at the bottom of the of the sight gauge and protrudes into the rounded portion of the glass tube.

The Raising Bubble Apparatus has a needle that is set and kept about 3–5 centimetres below the flat portion of the tube. The sight gauge and glass tube are prefilled with deionised water at the initial test pressure and reservoir temperature., the reservoir oil is then injected downwards into the flat glass tube, displacing the water until only the lower circular portion of the glass tube contains any water.

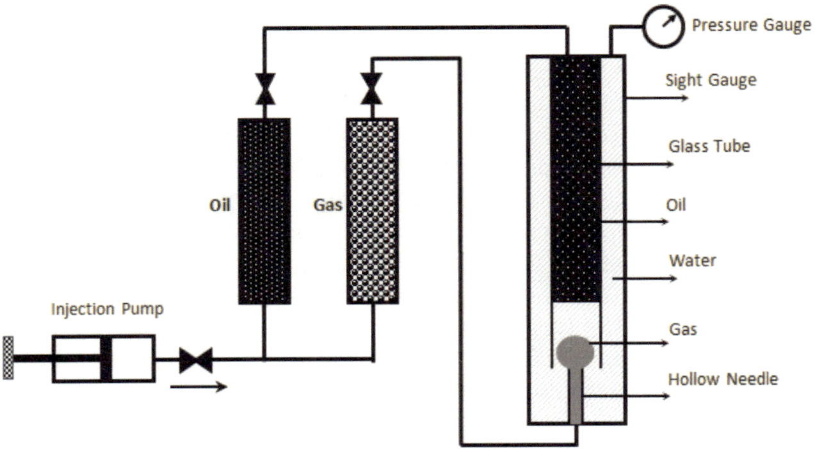

Figure 12.
Raising bubble apparatus by core lab [8].

A small gas bubble must be placed at the tip of the hollow needle and liberated into the tube, it will rise through the water, through the water/oil interface and up through the column of oil. After two or three bubbles have risen through the oil the water oil interface, then it is replaced with a fresh reservoir oil.

All the process of the gas bubble raising is monitored using a motion tracking optical system with a video camera mounted on a rail parallel to the path of the rising bubble. A magnified view of the bubble can be observed on screen and recorded as a small movie clip (mpeg). The time of the raising process for each injected gas bubble is calculated for each test pressure and use this data to interpolate the MMP.

12.4 Zero interfacial tension (VIT vanished interfacial tension)

At the Miscible Pressure, no interface exists between crude oil and injection fluid, i.e., interfacial tension approaches zero **Figure 13**. The VIT method measures the interfacial tension between the two phases, the measurements are done in a high-

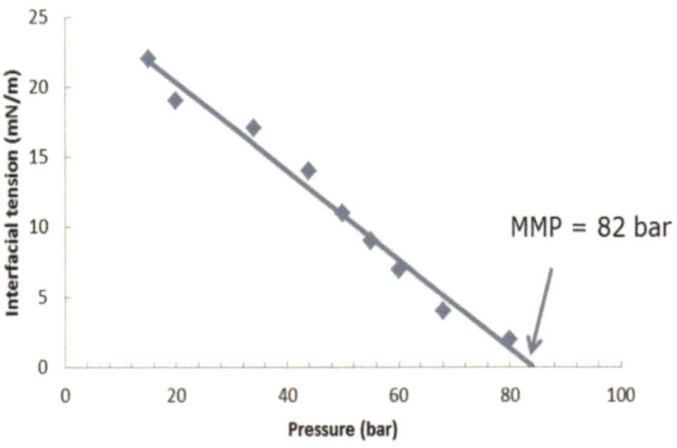

Figure 13.
Vanishing interfacial method to estimate the MMP.

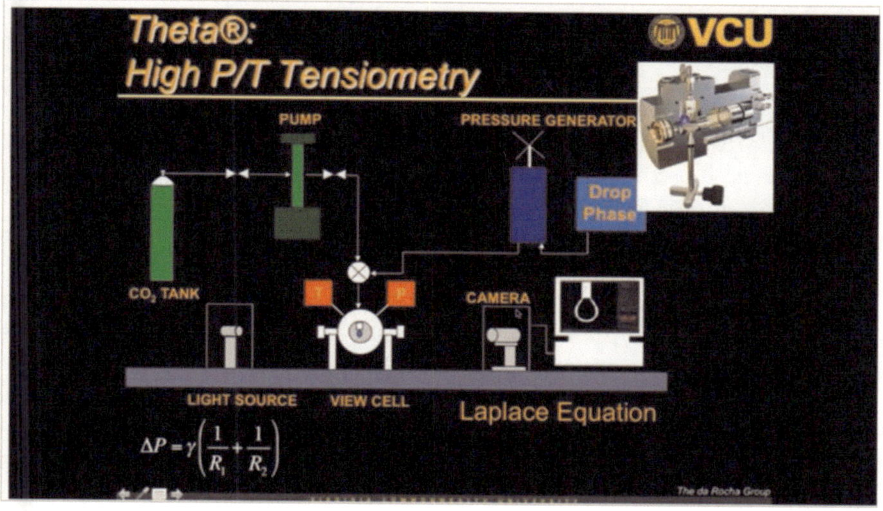

Figure 14.
High P/T tensiometer for the VIT [9].

pressure cell with an optical tensiometer, the oil is introduced as a drop phase into the chamber filled with the injection fluid [9].

The interfacial tension is measured with the Pendant Drop Shape analysis. It is measured at 5–10 different pressures at reservoir temperature after which the line is extrapolated to zero IFT. When the interface between two phases vanishes in all proportions, that is first contact miscibility. In a field CO_2 project, it is always multi contact miscible, never first contact miscible.

Although it has been established that VIT is NOT a rigorous measurement of MMP, it provides a good approximation. The measurement of MMPs in a high-Pressure Temperature Tensiometer apparatus **Figure 14** it can be done in two weeks for 20 ft columns and less than a month for 6-point, 80-foot columns [9].

12.5 Correlations to estimate the miscibility pressure

Stalkup, JR [10] presented his Correlation to estimate the MMP (1983-4) developed from 9 different miscible process displacing oil of different compositions with gas composed by more than 80% mol methane, it correlates MPP as function of the oil composition and saturation pressure. The correlation results exhibited average deviation of 260 psi, and maximum deviation of 640 psi; however, the correlation exhibited large errors for displacements with gases with methane content lower than 80 mole percent [11].

12.6 When is the best time to determine the minimum miscibility pressure?

The time when to determine the Minimum miscibility pressure is the nearest time to the reservoir discovery date, as soon as a PVT sample is available and analysed, other parameters need to be considered, as the reservoir dimensions, volumes of initial oil in places, the reservoir pressure, the reservoir conditions either undersaturated or saturated reservoir, recovery factors, among others.

A MMP process is not applicable for a saturated reservoir because saturated reservoirs are characterized for having initial pressure below the bubble point pressure evidenced by a primary gas cap.

To achieve a miscibility displacement in a saturated reservoir, it would be required to re-pressure the reservoir to above the bubble point pressure, which could be achieved injecting a considerable volume of gases or water while the reservoir is closed to producing, this situation is very improbable to happen because the prohibitive cost that represent injecting fluids without hydrocarbon production.

Under the current trends of switching the energy supply from fossil to renewable cleaner energies might be opportunities where the Carbon Capture and Storage (CCS) activities might supply the opportunities to use depleted oil reservoirs with low recovery factors for CO_2 storage. In those cases, dedicated geosciences and reservoir engineering studies will be required to mature every particular field case with two folds objectives, first to storage the CO_2, second to investigate how the remaining oil in the depleted structure will be affected by the injected CO_2.

For a subsaturated reservoir, the situation is different because there is higher probability to implement a miscible process with the purpose to increase the oil recovery factor and the reserves; the suggested procedure is:

1. Determine the initial Reservoir Pressure at datum, which is usually a horizontal line the gravity centre of the reservoir,

2. Determine the bubble point pressure at the reservoir datum.

3. Determine the operational pressure for the reservoir that satisfy the production objectives, considering the fluid composition, content, and onset of asphaltenes deposition, water injection ongoing operations.

4. Determine the difference between the initial reservoir pressure and the bubble point pressure, this difference determines the range of pressure available to deplete the reservoir before the start of the gas injection to achieve a miscible displacement.

5. Establish the ratio reservoir pressure depletion as function of extracted oil volumes.

6. Calculate the time to reach the bubble point pressure using the reservoir pressure depletion ratio as function of extracted oil volumes.

7. The time estimated from the above procedure will give insight of the time when initiate the gas injection to achieve a Miscible Displacement process.

This process must be based on numerical reservoir simulations using calibrated reservoir simulation models.

The MMP process are viable only when the incremental production generates enough revenue to cover the project cost implementation and generate revenues for the operator and the shareholders. In the field history cases we describe a case of multiple contact miscibility with dry gas in a subsaturated reservoir.

13. Field case: early identification of multiple contact miscibility injecting dry gas el furrial field, a case of multiple contact miscible gas injection combined with low-salt water injection EOR projects

El Furrial field is a giant structure discovered in 1986 by Lagoven s.a affiliated of PDVSA with the perforation of the Ful-1X well in Eastern Venezuela **Figure 15**, it encountered 854 m gross interval with 366 m net oil sandstone **Figure 16**, it was the more important discovery in South America in over 25 years, achieved 10 years after the nationalization in 1976, at a time when national production had declined to its lowest point since 1950; this field become the highest producing oil field in Venezuela, reaching a peak of 480,000 bbl/day in 1998. Secondary and Tertiary recovery studies to maximize the oil recovery were initiated in 1990 (**Figure 17**).

The discovery of this giant near 7.9 billion barrels of oil in place was a tremendous success of the exploration campaign undertaken by Lagoven s.a. [12] in an area where international operators exploited shallower oil reservoirs before the 1976 nationalization leaving unnoticed deeper structures containing a trend of giant light oil reservoirs.

Initially, a water injection project to inject 200,000 barrels of fresh water was designed to maintain the reservoir pressure at or above 6500 psi at reservoir datum, with the purpose to have a safety margin above the asphaltenes flocculation onset that was extensively measured to start at about 1500 psi above the bubble point pressure 4500 psi.

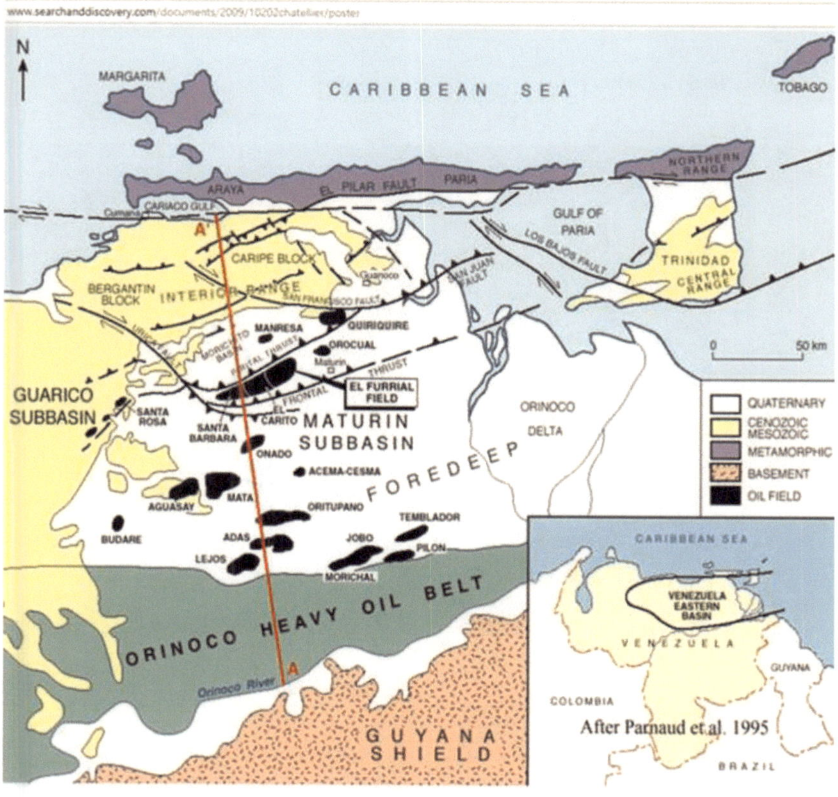

Figure 15.
El Furrial field location.

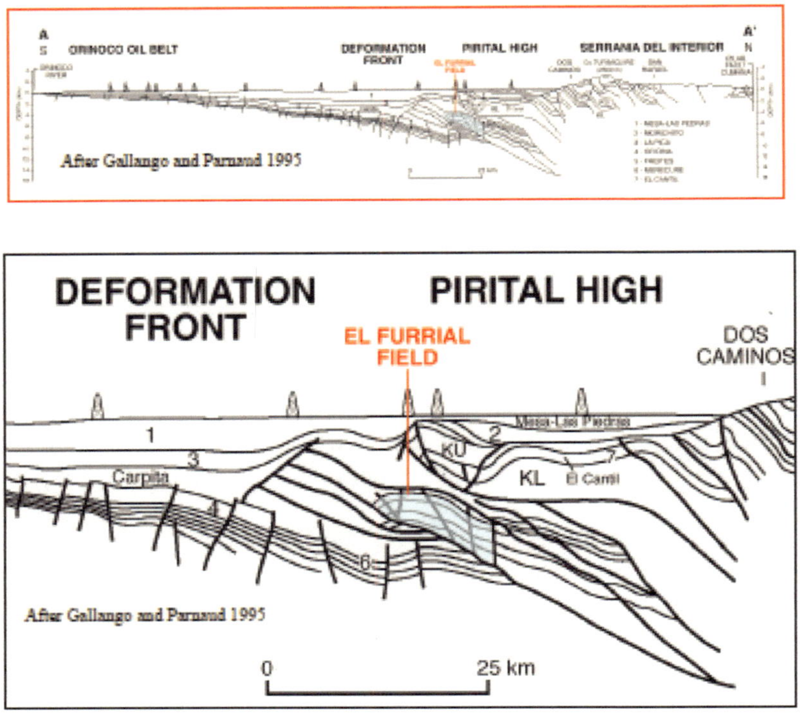

Figure 16.
Regional structural setting El Furrial field.

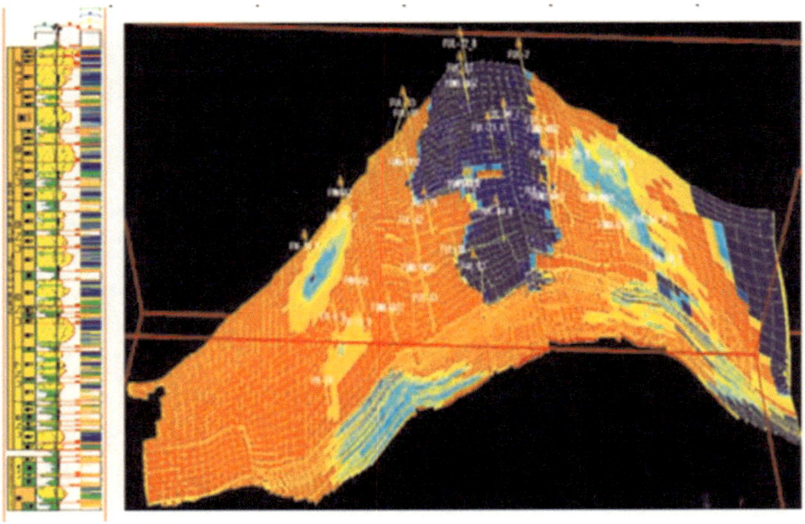

Figure 17.
Full field reservoir simulation model.

Immediately after the sanction of the water injection project, it was identified the feasibility of a Multiple Contact Miscible (MCM) process injecting Dry Gas [13]; the feasibility was identified from one-dimension compositional simulations, followed by experimental studies that concluded with the implementation of a project to inject

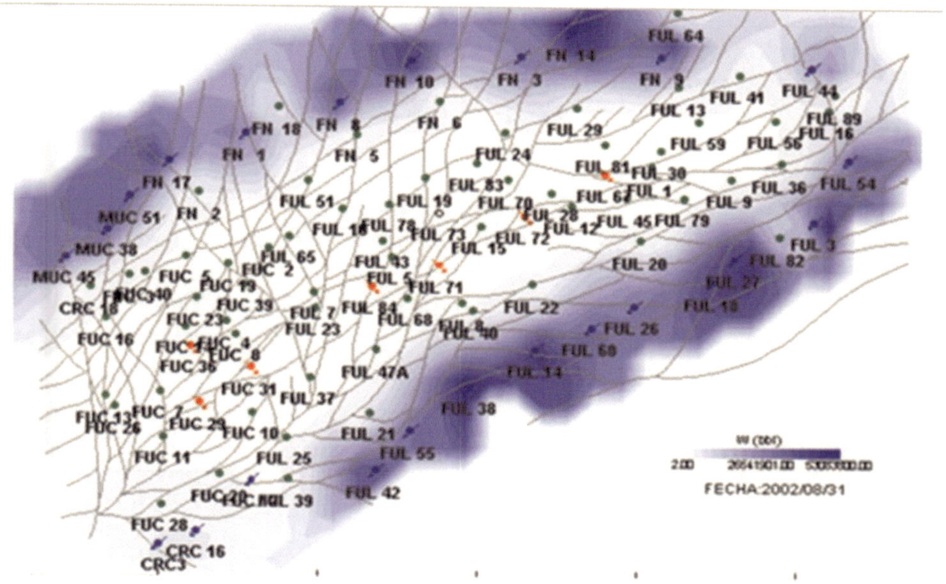

Figure 18.
Water and gas injection project and oil production wells map (IAGF project).

600 mmscf/d of dry gas to develop a MCM project in parallel to an increase of the water injection to 450,000 barrels/day to significantly increase the final recovery factor (**Figure 18**).

The two projects water injection and MCM gas injection were designed, planned and implemented to operate simultaneously. The gas injection in 5–6 wells located at the crest of the anticline and the water injection in near 36 water injection wells positioned at the flanks in two independent rows one for each main reservoir unit (**Figure 18**).

The projects were designed to achieve a combined recovery factor in range from 55 to 60%. Both projects were timely implemented and successfully operated since inception to mid project life. By end of 2021, 36 years after the field discovery the recovery factor achieved is estimated in range of 44.5% that is near 10–15% below the initial predictions; this implies that a range from 845 to 1240 million barrels were not produced.

The procedure applied to determine the Multiple Contact Miscibility displacement is summarized below:

1. A one dimension compositional model was set up to simulate the displacement of reservoir oil injecting dry gas, assuming all liquids would be extracted before the injection. The displacement was modelled at 6500 psia the minimum reservoir operational pressure predefined based on the asphaltene deposition experimental studies. The one-dimension model indicated a multiple contact miscible gas process would occur as a result of the thermodynamic effects injecting gas at pressures 2000 psi above the bubble point pressure.

2. The equilibrium constant at 6500 psi were experimentally determined by swelling tests performed in Core Laboratories in absence of correlation for pressures above 4000 psi.

3. A slim Tube Test apparatus was set up at Intevep Technology Centre, calibrated injecting soltrol, a high-purity, low-odour, low-toxicity synthetic solvent, in the porous media saturated with mineral oil at 6500 psi the reservoir operational pressure, obtaining a recovery close to 95% recovery, indicating the occurrence of miscibility.

4. The slim was saturated with crude oil samples taken from the Ful-12 well located at the crest of the structure, where the gas injection operation were planned.

5. The gas sample for the displacement in the slim tube was captured from the national gas network, because the injection displacement was planned with a dry gas; the analysed showed a minor percentage of CO_2.

6. The gas injection was executed at 6500 psi, and the recovery registered in range of 92–96% confirming multiple contact miscible displacement injecting dry gas at 6500 psi as observed in the one-dimension compositional model.

In this specific case, the slim tube tests were performed at 6500 psi to verify the occurrence of miscibility previously calculated in the one-dimension compositional simulations.

Afterwards the MMC miscible gas injection project known as IAGF (Injection de Agua y Gas Furrial) was sanctioned and implemented along with an increase in the water injection capacity; the gas injection started in 1998. The performance of the two projects is discussed below:

13.1 Field case IAGF gas injection combined with low-salt water injection El furrial production performance

The field production history of the IAGF project described in the previous section is shown in the **Figure 19**, the pressure history is presented in the **Figure 20**. The field production targets for the field were planned with the premise to maintain the reservoir pressure in range of 6500–7000 psia to ensure the gas displacement under the miscibility process and to avoid the asphaltene deposition around the wellbore of the producer wells, plugging the perforations in the well completions and damaging the formation in the wellbore zones. The asphaltene flocculation onset pressure was measured as function of the asphaltene in around 6000 psia.

The **Figure 20** shows pressure performance, it is observed the reservoir management activity to maintain the reservoir pressure in the predefined range 6500–7000 psia was consistent since the inception of the project until approximately the year 2008, when a drastic pressure decline occurred as a result of lack of continuity in the gas and water injection operations, that occurred in parallel with an intensive infill drilling campaign implemented in the lapse 2008–2010; such drilling campaign resulted in an increase of the production however the reservoir voidage was not maintained causing a drastic reduction of the reservoir pressure to levels below 6500 psia the operation pressure that preceded a dramatic fall in the production rates observed in the years 2013–14.

The analysis of the historic performance suggests this reservoir was capable to produce at rates higher than predicted for a longer period (after 2004), if the reservoir pressure would be maintained above 6500 psi.

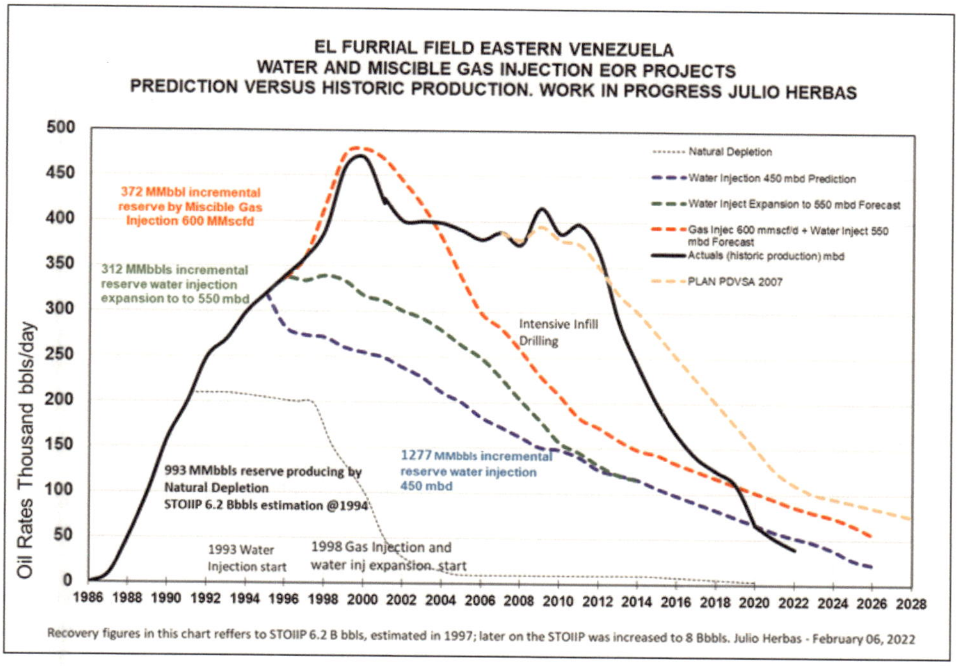

Figure 19.
Production forecast plot IAGF project.

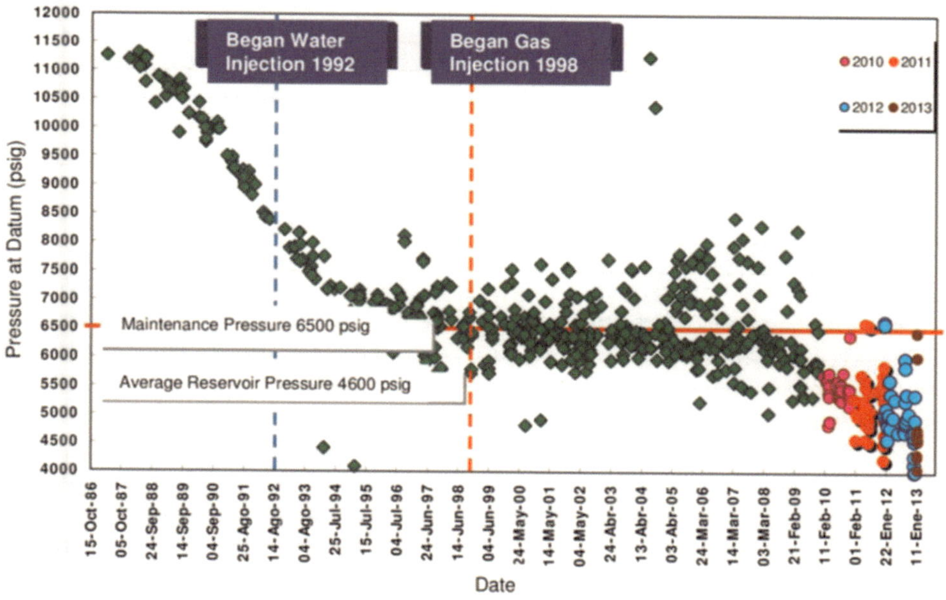

Figure 20.
Pressure performance plot IAGF project.

The project management seems to have been disattended in the last 10 to 12 years because of the nationalization of the injection facilities, politization of the project management, and other detrimental practices of century 21 socialism regime, as a result the reservoir pressure declined to 4600 psi in 2013, below the asphaltenes flocculation onset.

At current standards of the technology with the application of best reservoir management practices the final recovery for this project should have been close to 70–75%; however, several events that affected the continuity of the gas and the water injection affected adversely its performance. Analysis of the results of the two consecutive EOR projects, are discussed below to illustrate the impact of early EOR gas and water injection studies.

The first project, a Low Salinity Water injection was designed in 1990–1991 to maintain the reservoir pressure at 6500 psi, the water injection was initiated in 1992 with a pilot test followed by the construction of the facility to inject 450,000 bbls/day of water to recover an estimate of 1277 million barrels of incremental oil (~20 % recovery above the primary recovery factor estimated in ~16%), it was named the Resor project. The water injection operations were initiated at reservoir pressure of near 8000 psia at 14,000 feet deep, fresh water of ~1000 ppm from shallow aquifers was selected after screening all available sources which put this project as a Low Salinity Water Injection displacement process.

Immediately after the RESOR project was sanctioned in 1992, the gas injection feasibility studies were initiated, the experimental and engineering work were executed in 2 years, and the obtained results concluded the dry gas injection would add an important increase in the recovery, therefore it was recommended to initiate a dry gas injection with pure methane at rates of 550 mmscf/day in the crest of the structure to increase the final recovery to around 55–60%. The Miscible Displacement Injecting Dry Gas studies were done with TCA Reservoir Engineering Services a company based in Durango Colorado with cooperation from the EOR department of the Texas University and the Research Institute Intevep S.A. affiliated of PDVSA; the simulation work included the one-dimension and full field compositional simulations that demonstrated a multiple contact miscible displacement injecting pure methane, this process was corroborated with slim tube experiments executed at Intevep the Technological branch of PDVSA, and verified in Core Lab and Westport Laboratories in Houston Texas, giving the bases for the Miscible Gas Injection EOR project initiated in 1998 (IAGF).

The Miscible Gas Injection project was designed in 1994 [13] to inject 550 million standard cubic feet per day of dry gas together with an expansion of the water injection to 550,000 bbls per day, with the objective to generate reserve of 684 million barrels of incremental oil, additional to the base water injection estimated reserve in 1277 million bbls; the gas injection project was named IAGF and initiated in 1998. The project aimed to achieve a multiple contact miscible displacement injecting dry gas at 6500 psia at the top of the reservoir, with the objective to remove and displace the residual oil toward the producer wells acting in combination with the water injection at the flanks of the structure, it implied the conversion of five-six producer crestal wells to gas injectors sacrificing a substantial oil production rate (**Figure 21**).

The cross section of the compositional simulations shows the effect of the miscible gas displacement in the reservoir reducing the oil saturation to near zero in the surrounding zones to the gas injection wells, as it was observed in the slim tube tests and predicted in one-dimension compositional models.

13.2 IAGF water and gas injection project, performance review

In this section, it is described an analysis of the actual reservoir response under water and gas injection versus the predicted forecasts, to illustrate the value of the

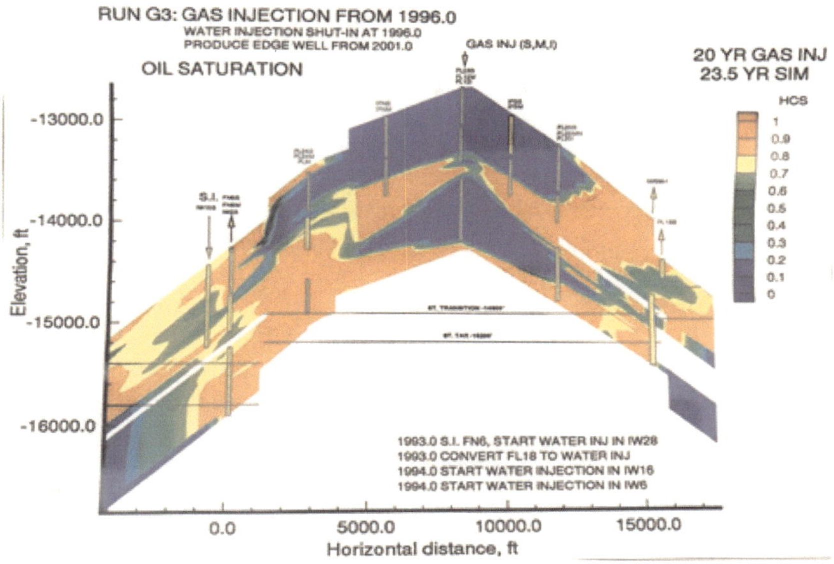

Figure 21.
Cross section compositional simulation, showing near zero oil saturation in zones affected by the miscible displacement injecting dry gas.

early identification and implementation of the feasibility of a MCM process, for this analysis, data from various sources was used to generate an overview of the project performance.

Both injection projects are classified as Enhanced Oil Recovery, because their intrinsic characteristics:

- The water injection at reservoir pressure of near 7500 psia at 14,000 feet deep, designed to inject fresh water places this project within Low Salinity Water Injection Displacement process.
- The Multiple Contact Miscible Gas Injection Process designed to achieve miscibility injecting dry gas at 7000 psia at the crest of the reservoir, to remove and displace residual oil toward the producer wells achieving a near zero residual oil saturation in the zones affected by the MCM displacement, acting in combination with the water injection at the flanks of the reservoir.

To evaluate the results of both projects, it was accessed the original working files and the numerical simulation models build in 1990–1997 built in house using commercial simulators with data generated in laboratories of Intevep s.a., Core Laboratories Inc, Schlumberger, and other international laboratories, as compatibility tests performed in Serk Baker labs in United Kingdom. We also used the universities Simon Bolivar and Universidad Los Andes in Venezuela, Bristol University in UK and Texas A&M University in Houston Tx. Before the project sanction, all the tests were verified by international laboratories in Houston and Dallas Texas USA.

The analysis of the production performance is based in the production forecast, actuals production profiles obtained from public domain literature, personal experiences, notes, and testimony of some of the main players.

The simulation profiles generated for the sanction of the two analysed projects were plotted along with a natural depletion case, and the actual historic production performance plots, shown in **Figure 19**, it shows the natural depletion and three consecutive cases corresponding to: injection water 450,000 bbls/day base case (blue dotted line), water expansion to inject 550,000 bbls/day of water (green dotted line) and the miscible gas injection of 550 mmscf/day combined with 550,000 of water injection.

The solid black line represents the actual historic production, the orange line starting at 2007 represents the PDVSA Plan at that year; the STOIIP at this year was increased to order of 7.9 B bbls.

A reservoir pressure review to the available data in the **Figure 20**, shows observed pressures in the first 1–2 years after the gas injection started in 1997, the actual oil rate was lower than predicted, it declines to 400,000 bbls/d and is maintained until the year 2012, then a dramatic decline started to reach the actual rate ~60,000 bbls/d in the year 2020–21.

The overall actual production performance is superior to the forecast because of intensive infill drilling; however, the dramatic production decline started in 2012 is result of a progressive discontinuity of the gas and water injection operations.

The main conclusion is the water and gas injection operations affected favourably the production overachieving the initial forecast and generating more than planned reserves, however a more careful analysis shows the achieved actual recovery factor just reach around 45% of the updated STOIIP (7.9 MMbbl) which is near 10% bello the recovery factor obtained in the initial predictions; this deficit is a result of discontinuation of the gas and water injection operations and the over-production above the established production levels, noticeably the dramatic production decline occurred in 2012, coincides with the pressure depletion below the operational maintenance pressure defined for the reservoirs in the implementation studies in 6500–7000 psi at datum.

Currently the gas flare is common in this giant field, which is a result of the inefficient operation in the compression and injection system; the field reservoir pressure was well managed until 2012, however after 2013 the decline in the reservoir pressure coinciding with the low oil production rates reflect its effects of well productivity deterioration (**Figure 22**).

14. CO_2 injection as EOR process

EOR CO_2 flooding consists of injecting large quantities of CO_2 in the reservoir to form a miscible flood, the injected CO_2 volume is determined from experimental and

Figure 22.
Press Note of Gas Flaring in El Furrial Field Monagas State, March 2020.

compositional studies can be from 15% or more up to 1.5 hydrocarbon pore volumes. When the CO_2 gets in contact with the oil in the reservoir, if the pressure is high enough, there will be a kind of "vaporizing gas drive" recovery mechanism:

1. The CO_2 extract the light–to-intermediate components from the contacted oil,
2. the miscibility is developed in the displacement front,
3. in the displacement front the oil viscosity is reduced when the CO_2 swells the residual oil, and
4. the mobility ratio will improve because of the oil viscosity reduction.

The CO_2 is an efficient miscible displacement solvent, it requires lower pressures to achieve miscibility compared to other gases as hydrocarbon gases, Nitrogen and flue gas, the CO_2 injection as EOR method in the oil industry is well known and has been applied in many fields.

Conversely in some fields with high CO_2 content, meaning the CO_2 is available, the CO_2 reinjection has not been implemented, the produced CO_2 is vented, or the wells have been shut in, in contrast with other fields and reservoirs where the CO_2 is not available and the required volumes for CO_2 injection for EOR purposes, had been purchased and transported from other sources. Those fields with high CO_2 content that have been under exploited because its high CO_2 content can be an excellent fit of the technology to produce the hydrocarbons with high CO_2, separate it and reinject in the reservoir displacing the total usable hydrocarbons and leaving the CO_2 reinjected volumes in the reservoir, finally used as the CO_2 captured recipient.

The **Figure 23**, is a schematic of a CO_2 EOR process followed by a waterflooding displacement to chase the CO_2 slug.

The Screening Parameters for a CO_2 EOR are listed below:

- Gravity >27 API Viscosity <10cp
- Composition C5–C20 (C5–C12) oil saturation >30% PV
- Formation type sandstone/carbonate Net thickness relatively thin
- Average permeability not critical
- Transmissibility not critical
- Depth >2300 ft
- Temperature < °250

14.1 EOR and CCS

Since the 1950s, the oil and gas industry has spent many billions of dollars on CO_2 EOR technologies, commercial projects, and developing operational knowledge. Most of this activity has been in land-based oil and gas fields. The first patent for CO_2 EOR was granted in 1952, the Texas Railroad Commission reports the first three projects were initiated in Osage County, Oklahoma between 1958 and 1962. These CO_2 EOR

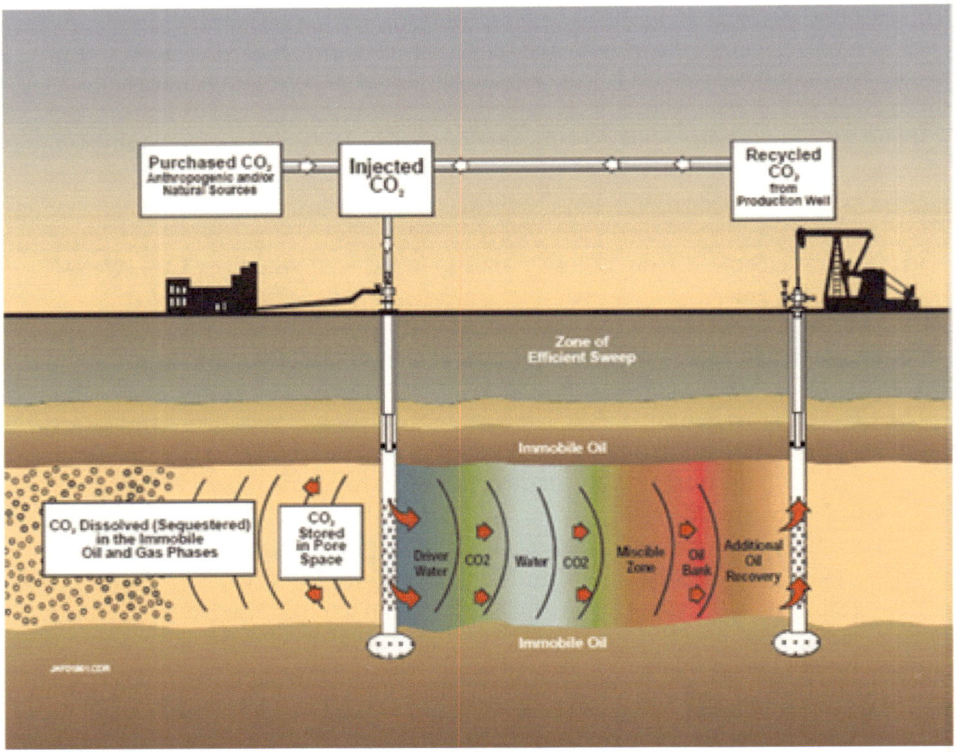

Figure 23.
Schematic of a CO2 EOR Displacement Followed by a Waterflood.

projects have steadily increased over the years based on the growing availability of CO_2 and technology advances.

In 2012 the Oil & Gas Journal EOR survey, reported the CO_2 flooding in the USA was producing more oil than EOR by Steam Injection (308,564 bbls/day vs. 300,762 bbls/day) with 41% of the output from all types of EOR. The active CO_2 EOR projects in the USA were increased to 120, representing 89% of the total 135 CO_2 EOR project globally. Furthermore, in the past years, as sources of CO_2 offshore and deep-water technology has become available new EOR CO_2 injection projects were initiated, as in the giant Lula field located deep-water in Brazil, which is the pointy end of a very long and successful industry history of CO_2 EOR.

Conversely, some onshore fields with large volumes of liquid CO_2, has not been fully developed because the high CO_2 content, which might have been produced and used to recover the hydrocarbons and leave the CO_2 storage in the reservoir.

15. Challenges and solutions injecting CO_2 for EOR

Some of the challenges for the EOR CO_2 project implementation are:

- CO_2 availability,
- Early breakthrough of CO_2 because the very low viscosity of the CO_2 results in poor mobility control, this is associated with the pressure insufficient for the CO_2 to swell and remain in the oil

- Corrosion in the producing wells
- Separating the CO_2 from saleable hydrocarbons,
- Repressuring of CO_2 for recycling, and
- Requirement of large volumes of CO_2 per incremental barrel produced.

Over the last 25 years, a small number of offshore saline aquifers and oil and gas reservoirs have successfully used many of the technologies developed through the last 58 years of land-based CO_2 EOR experience. It is possible that CO_2 is a viable means to increase hydrocarbon output from many depleted offshore reservoirs that are marginal or no longer productive; most operators are not using this technique on their reservoirs because they do not have an economical supply of CO_2, other operators because do not own the CO_2 technology. However, cost-effective supplies of CO for many of these offshore fields may become available as carbon capture from nearby electric power plants and other large, stationary sources of CO emissions becomes more common (**Figure 24**).

In the last years, it has been detected some onshore light waxy oil fields containing large amounts of CO_2, one of those fields located in Europe and another one in South America, both were not fully developed because the high CO_2 and wax content has somehow affected the wells productivity, so that their achieved recovery factor has been very low near 1% of the initial oil in place; their production operations were affected by other factors as pressure depletion, while having a large volume of liquid CO_2 dissolved in the gas cap, those CO_2 volumes instead being flared, might be compressed and injected back into the reservoir.

Challenges for offshore EOR CO_2 injection and CCS projects are more stringent because the higher development costs, the offshore surface facility space, weight and power limitations, the lack of sufficient and economical CO_2 supplies, and fewer existing wells that are more widely spaced. All these factors are added complexity that contribute to uncertain EOR performance and require longer time periods for CO_2

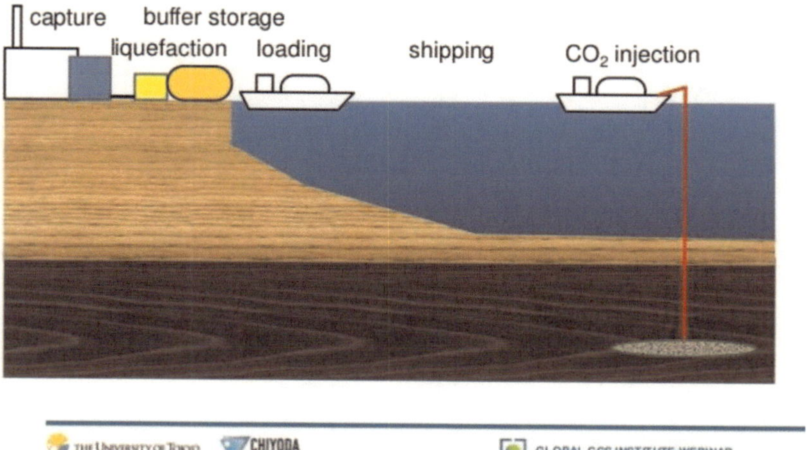

Figure 24.
Schematic CO_2 capture and injection offshore.

placement to displace oil and gas and achieve adequate sweep efficiency. However, EOR is currently being considered for several offshore developments. The prognosis is better when successful secondary recovery methods have been employed through water and natural gas injection, which make CCS and CO_2 EOR methods much more feasible and less costly to apply.

Some of the key challenges and solutions for offshore CO_2 injection for EOR and CCS projects, include the use of CCS tanker ships and barges to ensure CO_2 supplies and to provide service facilities until the construction of pipelines and construction of permanent facilities is justified. Horizontal well designs may be needed to offset a lower well density and achieve a more uniform sweep and displacement.

Transport of CO_2 from onshore sources to offshore oil and gas fields has been successfully done at several CO EOR projects using pipelines and barges. Tanker ships have successfully and safely transported CO_2 for over twenty years, are best suited for the small volumes needed for pilot CO_2 injection tests; tanker ships that deliver LNG

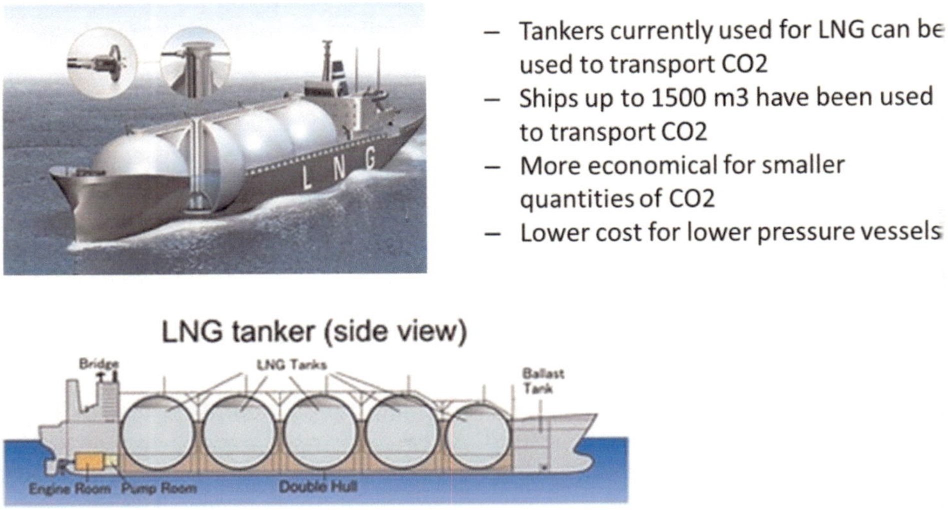

Figure 25.
CO2 transport in LNG tankers.

Figure 26.
LNG Tanker ships for CO2 Delivery Offshore EOR projects (OTC 21984).

to ports with supplies could carry CO_2 it on their return voyages to economically supply EOR projects (**Figures 25** and **26**).

Some CO_2 activities that are happening in the world are the CO_2 pipeline projects in planning or in construction in several continents, as:

1. A pipeline planned between Mississippi and Texas to supply EOR projects with CO_2 from anthropogenic and natural sources,

2. Two offshore projects for permanent storage of CO, offshore Norway, the Sleipner and Snøhvit CCS projects.

There are at least three CO_2 EOR and enhanced gas recovery (EGR) projects around the world, as:

1. Gas injection in the Bay St. Elaine oil field in the Louisiana marshlands,

2. Dulang field WAG project in Malaysia's east coast in the South China Sea.,

3. Lula CO_2 offshore Brazil

16. CO_2 capture and injection, costs, and technology

A critical element of a CO_2 capture and storage project is obtaining the CO_2, the technology for separating it from a flue gas, and the business model of who pays, it has been surprising to hear discussions about where the CO_2 is going to come from for capture and storage purposes, and its associated costs, the oil operators being obliged to do carbon capture and storage will look for the cheapest way to obtain CO_2 which would otherwise be vented to the atmosphere. I obvious should be to reinject the captured CO_2 into the depleted oil reservoirs, which should increase the reservoir pressure and in cases helps to increase the oil production and final recovery factors [14, 15].

For the natural CO_2 subsurface reservoirs, the CO_2 is already in the ground, it is not sensible to produce it to storage unless a reasonable use is determined, that can be EOR to increase the oil recovery. Gas wells with high CO_2 should be studied, although they would normally not be produced at all, an option can be to capture the CO_2 and reinject it to enhance the oil recovery and after extracting it, use the reservoir to storage the CO_2.

Flue gases from power stations and energy intensive industries, particularly concrete, steel and oil refining are potential sources. The original assumption for carbon capture and storage was targeted to coal power station flue gases. This proposition in the UK and Netherlands now have the expectation to stop using coal power, which will reduce an important source of CO emission. The UK anticipates continuing to use gas power, the Netherlands anticipates all power generation coming from renewables [16].

The focus to flue gases from energy intensive industries with typically 20 per cent CO_2, that needs to be separated from the 80 per cent of other gases, the amine technology separates the amine molecule attaches to the CO_2 in one column, and the amine is separated from the CO_2 in a second column, cost estimated were reported in range \$35 to \$69 per tonne of CO_2 captured from a coal power flue gas in India by 2019 [17].

The key measures to report cost of the CCS defined by the Global CCS Institute in 2017 update, defines it as the life cycle unit cost of production and cost per tonne of avoided CO_2. The cost per tonne of CO_2 avoided is a measure that enables comparison across various technology in terms of their value reducing greenhouse gas emissions. The costs for USA reported in 2017 show for flue gas from cement in cost in range 53 to101 US$/tonne; for Iron and steel the cost is reported in 95 to 370 US$/tonne [17].

There have been many efforts over the past 10 years or so to find ways to reduce these costs. One idea is for fuel to be combusted in pure oxygen, with an air separation upstream of the combustion unit, a mature technology, then the flue gas is near entirely CO_2, this option has been studied by Occidental (OXY) (**Figure 27**).

A great deal of research is going into carbon capture technology, particularly with new solvents. An example is the advances from Occidental Petroleum (OXY) to direct capture and storage CO_2 from the air (DAC): [18].

In 2019, OXY Low Carbon Ventures (OLCV) released a first look of design of the plant to capture up 500 Kt of CO_2 annually directly from air to be used in EOR projects and subsequently stored underground permanently in the Permian Basin, expected to expand to include multiple DAC plants, each capable of capturing one megaton of atmospheric CO_2 annually. If the initial plant is approved by Occidental and Carbon Engineering, construction is expected to begin in 2021, with the plant becoming operational within approximately two years.

On March 28, 2022, Oxy subsidiary (OLCV) and Weyerhaeuser Company (WY) announced an agreement for the evaluation and potential development of a carbon capture and sequestration project in Livingston Parish, Louisiana. The agreement provides OLCV with exclusive rights to develop and operate a carbon sequestration hub on more than 30,000 acres of subsurface pore space controlled by Weyerhaeuser. OLCV will use the land to permanently sequester industrial carbon dioxide (CO_2) in underground geologic formations not associated with oil and gas production, while Weyerhaeuser continues to manage the aboveground acreage as a working forest [18] (**Figure 28**).

Figure 27.
Tanker Transporting Oil Cargos a Sources of CO Discharge to the atmosphere.

Credit: Carbon Engineering.
This rendering shows a first look of what will be the world's largest direct air capture facility, currently being engineered by Carbon Engineering and 1PointFive.

Figure 28.
DAC direct air capture CCA oxy plant.

The agreement, is a pivotal step in OLCV subsidiary 1PointFive's strategic vision to develop a series of carbon capture and sequestration hubs within the U.S.:

- 1PointFive plans to build, acquire, and operate multiple sequestration hubs on the Gulf Coast and across the U.S., some of which are expected to be anchored by Direct Air Capture (DAC) facilities, to offer storage capacity to point-source emitters, such as manufacturing sites and power plants, with a capacity to sequester up to hundreds of millions of metric tons of anthropogenic CO_2.

- 1PointFive aims to play a transformational role in combatting climate change through industrial decarbonization of the hard-to-abate industrial sector in the U.S.

17. Who pays the CO_2 capture costs?

The question of how to cover the costs of CCS are being discussed for several years, one idea has been the CO_2 utilization using electricity to make hydrogen and using this hydrogen for CO_2 activation towards methanol or methane.

In Europe, emitters of CO_2 are being hit by ever increasing regulatory pressure and costs to dis-incentivize emitting CO_2 to the atmosphere, the costs of CO_2 separation from flue gases would be paid for by emitters. This is the plan of the Rotterdam PORTHOS project, which envisages that energy intensive industries in the Port of Rotterdam would pay themselves for CO_2 capture and storage.

In Europe, the emissions trading scheme covers all land-based emissions (not shipping and aviation), the cost of emitting is not yet close to the cost of CO_2 capture and storage, and is not a stable price, so does not provide enough incentive by itself. The Netherlands and UK Governments are looking provide a subsidy or additional tax, between the carbon price and the cost of carbon capture.

18. CO_2 to enhance the oil production

Research work conducted in slim tube test experiments to investigate the recovery with injection of various gases, such as CO_2, N_2, CH_4, or flue gas, have demonstrated the CO_2 gas injection results in the highest oil recovery factor compared with injection of the other gases.

In water flood secondary recovery projects, after long periods of water injection, a significant amount of oil remains in the reservoir due to the capillary pressure between water and oil. In these cases, the oil recovery can significantly be improved by gas injection in such a way that the gas/oil front moves gravity stable through the reservoir.

If the injected gas and the displaced fluids are moving in gravity stable displacement, substantial incremental oil can be produced; the factors driving the incremental oil production are, reduced interfacial tension for miscible or near miscible displacements at reservoir conditions, gravity drainage for injection of non-miscible gases and improved sweep efficiency for attic oil with stable front moving vertically through the reservoir.

Several gases can be injected in the reservoirs, for a case when methane, CO_2 and N_2 are available, the choice of what gas will be injected depends on the prices of the gases, costs of injection and incremental oil recovery by the respective gas.

18.1 Outlook and growth potential

The current world energy market trend driving the transition to clean energies to replace fossil energies and reduce the CO_2 emissions, implies a progressive reduction of the oil and gas production; at this point some industry observers believe the EOR methods may help offset the predicted decline in oil production over the next twenty years, and the CO_2 EOR may be a substantial portion of the future EOR growth. A key factor for this growth is a sustainable economic supply of technologies and CO_2 for injection where the CCS initiatives might be an important factor.

19. CO_2 EOR field cases Lula Project Brazil

The Lula field is a supergiant ultra-deep water offshore field located in the Santos basin southeast of Brazil, it is the most significant CCS project in Brazil. And Latin America, it is a pre-salt carbonate reservoir in the Santos Basin located below a thick, 2000 m salt column trapping a light, 28–30° API oil with high solution gas ratio (200–300 m^3/m^3 and variable CO_2 content between 1 to 15%, with neighboured areas with up to 80% of CO_2. The Lula field was developed in phases in the prospective areas of the field defined with extended well tests, production pilots followed by large scale production developments. The pilots provided data to calibrate simulation models, select strategies to maximize recovery and profitability, for the development of other fields in the Santos basin pre-salt blocks.

Early studies showed the oil recovery factor could be greatly improved with secondary and tertiary recovery by implementing a Water-alternating-gas (WAG) injection EOR project chosen because of the availability of seawater, produced gas, and the reservoir conditions particularly suited to miscible methods mixing water and gas. The project that began in 2009 with the arrival of a floating, production, storage, and

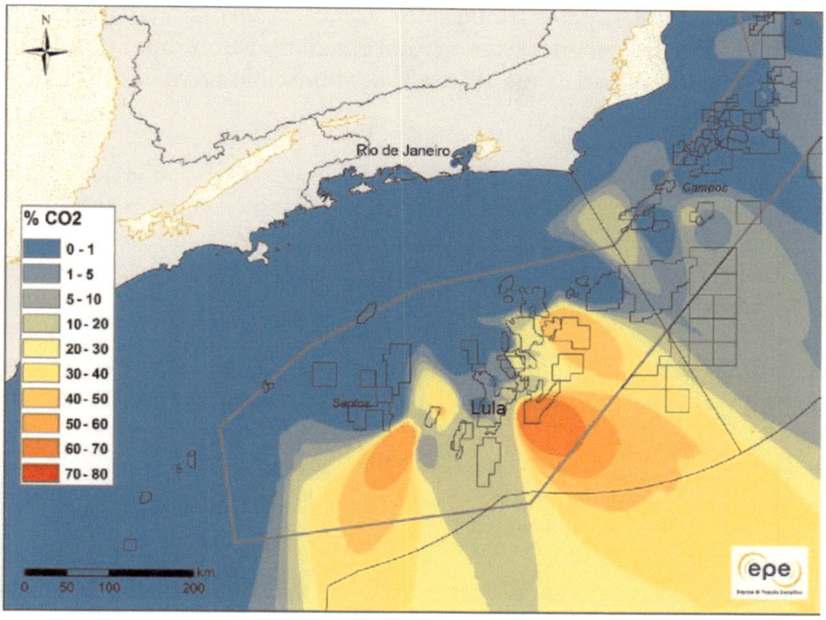

Figure 29.
CO2 Concentration in the Eastern Bank of Brazil's Exploration Areas. Source: EPE Empresa de Pesquisa Energetica January 2020.

offloading vessel, followed by the WAG pilot project with three producer wells with one gas injector at about 1.0 million m^3/day initiated in April 2011, the facility began exporting some gas to shore and the injection wells began to inject mostly CO_2 at rates of about 35000 m^3/day. The pilot was monitored with permanent downhole pressure gauges, and gas injection tracers.

The first results were presented in the SPE155665, concluding the injection of WAG using CO_2 separated from the associated gas in the pilot project as a suitable strategy to increase the oil recovery; the 2012 production and pressure data monitoring of the WAG installation was translated into EOR expansion at the field scale. By December 2018, there were nine production systems (FPSOs) at Santos Basin, with natural gas pre-treatment and CO_2 separation systems using membranes. Since 2013, up to end 2018 around 9.8 million metric tons of CO_2 has been injected, and the projects continue in operation.

This CO_2 pilot project made the Lula field a pioneer in Deepwater CO EOR, Petrobras may set the record as the first company to successfully combine CCS and CO_2 EOR for large-scale, sustained oil production in deep-water (**Figure 29**).

20. Conclusions

1. The understanding of the feasibility to develop a Miscibility Displacement process at early stages of a field development along with the implementation and economic studies is crucial for any relatively large light oil reservoir.

2. A surveillance program to determine the pressure decline trend as function of produced oil and a sound fluid characterization are paramount to establish an optimum operational pressure aiming to achieve high oil recovery factors.

3. The typical candidate gases for injection are dry or wet natural gases, Nitrogen, CO_2 and flue gas, a product from natural gas combustion; from those gases, the CO_2 has been identified as the more efficient miscible agent based in its property to dissolve the oil.

4. The current trends of Carbon Capture and Storage (CCS) objectives pursued by the industry to reduce the green house effects can be levered with the implementation of more CO_2 miscible injection projects elsewhere the CO_2 is available, as there are several oil fields that have not developed because the hight CO_2 content.

5. The gas injection operations efficiently planned, engineered, implemented, and operated with rational criteria contributes to the reduction of CO_2 emissions.

6. The field case of a multiple contact miscible gas injection project in Venezuela is an example of value destruction by political intervention that has resulted in a large source of CO_2 emissions.

7. The actual intensive trends in Carbon Capture and reductions of CO_2 emissions worldwide are affecting the research and development activity in the oil and gas industry.

8. Brazil has more than 25 years' experience in carbon dioxide injection for EOR operations, which provides significant experience in the technology for CCS development and deployment. The Energy and Carbon Storage Research Centre in Brazil opened in 2007 promote and explore ways to make CCS commercially viable.

Acknowledgements

Credits for the RESOR and IAGF water and miscible gas injection projects

The implementation of the RESOR water Injection project was initiative of Mss Juana Albornoz (deceased) supported by Dr J.P. Chalot and other Lagoven's professionals, the RESOR team was composed by: Orlando Dumont, Jesus Nunez, Richard Alvarez, Magno Romero, Pascual Marques and Jesus Tineo, the Reservoir Engineering Team was composed by Raul Mengual, Antonio Russo, Luis Ruiz, Jose Gil, Nidia Pinto, Pablo Saavedra, and other professional. In the construction of the simulation models, it was notable the contribution of Dr Pepe Bashbush, and Richard Smith from Intera SRL.

The discovery of the multiple contact miscible injecting dry gas in El Furrial Field was done by Dr Elmond Claridge (deceased) Head of EOR Dept at the University of Houston, Dr Michael Todd and Dr Curtis Chase (deceased) both from TCA reservoir engineering services; the direction of the research project by Julio Herbas, supported by an extended team composed by geoscientist and engineers from Lagoven and Intevep s.a., and audited by a BP Exploration team: Ian Roberts, Dr Shen Tai Lee (deceased), Dr Mike Christie (fluids behaviour), Dr Neville Jones (Geoscientist), Mike Levitan (reservoir simulation), Andy Johnston (Compression Design), and Leon Miura from Lagoven s.a. in charge of the project implementation.

Conflict of interest

The authors wish to confirm that there are no known conflicts of interests associated with this publication and there has been no financial support for this work that could have influenced its outcome.

Acronyms and abbreviations

bbls/day	barrels per day
bbls	barrels
CO_2	Carbon Dioxide
CCS	Carbon capture and Storage
CO	Carbon
C1	Methane CH_4
C2	Ethane C_2H_2
C3	Propane C_3H_8
C4	Butane C_4H_{10}
C5	Pentane C_5H_{12}
C6	Hexane C_6H_{14}
C7	Heptane C_7H_{16}
EOR	Enhanced Oil Recovery
EOS	Equation of State
F	Fahrenheit
FCM	First Contact Miscibility
IAGF	Inyección de Agua y Gas Furrial
MMP	Minimum Miscibility Pressure
mmscf/day	million standard cubic feet per day
M^3/m^3	cubic meter/cubic meter
N_2	Nitrogen
MCM	Multiple Contact Miscibility
Psia	Pounds per square inch absolute
ppm	parts per million
PVT	Pressure Volume Temperature
Psi	Pounds per square inch
P	Pressure
STOIIP	Stock Tank Oil Initial in Place
T	Temperature

Author details

Julio Gonzalo A. Herbas Pizarro
Mineaoil Limited, London, United Kingdom

*Address all correspondence to: jherbas@mineaoil.com; julioherbas@btinternet.com

References

[1] Mansour EM, Al-Sabagh AM, Desouky SM, Zawawy FM, Ramzi MR. Experimental approach of minimum miscibility pressure for CO_2 miscible flooding: Application to Egyptian oil fields. International Journal of New Technology and Research (IJNTR). 2016;2:105-112

[2] Trent Jacobs Journal of Petroleum Technology. 1 February 2021. Available from: http://www.marinetechnologynews.com)/blogs/water-alternating-gas-injection-700522SPE-199121-MS (6)

[3] Perm Inc TIPM Laboratory. Fundamentals of Flow in Porous Media. Slim Tube Test for Determination of Miscibility Condition | Fundamentals of Fluid Flow in Porous Media. Available from: https://www.perminc.com/

[4] Chuck Kossack Schlumberger PUT (Petroleum University Technology) EOR. 2013

[5] Taber JJ, Martin FD, Seright RS. EOR Screening Criteria Revisited—Part 1: Introduction to Screening Criteria and Enhanced Recovery Field Projects. New Mexico Petroleum Recovery Research Centre; 1997

[6] Aladasani A. Updated EOR screening criteria and modeling the impacts of water salinity changes on oil recovery [doctoral dissertations]. 2012

[7] Interface Fluidics. Available from: https://www.interfacefluidics.com/pvt-and-fluid-phase-behaviour/

[8] Core Laboratories. Available from: https://www.corelab.com/ps/minimum-miscibility-rising-bubble

[9] Laurent S. Vanishing interfacial tension measurement—A fast route to minimum miscibility pressure. Biolin Scientific; December 2016. Available from: https://biolinscientific.com/

[10] Stalkup FI Jr. Miscible displacement. SPE Monographe 1984;8:13

[11] Ali Danesh A. PVT and Phase Behaviour of Petroleum Reservoir. Edinburgh: Heriot Watt University; 1997

[12] Chatellier JY, Olave S. Successful Exploration in a Trust Belt, Lessons Learned from the Giant Fields of Eastern Venezuela (the Furrial Trend). College Station, TX: Tecto Sedi Integrated Inc., Calgary Texas A&M University

[13] Todd M, Claridge E, Chase C, Herbas J, Marquez P. Milton Mendes Lagoven S. A., Carlos Espinoza Intevep S.A. Preliminary Investigation of Enhanced Oil Recovery by Gas Injection in the Furrial Reservoir. TCA Reservoir Engineering Services; December 1992

[14] Ghorbani M, Gandomkar A, Montazeri G. Describing a strategy to estimate the CO_2 heavy oil minimum miscibility pressure based on the experimental methods. Energy Sources, Part A: Recovery, Utilization and Environmental Effects 2019;41(17):1-11

[15] Global Costs of Capture and Storage, Global CCS Institute 2017 Update

[16] Negrais Seabra P. Lula Oil Field CO_2-EOR Project Update. 4th International Workshop on Offshore Geologic CO_2 Storage – 11. Norway: University of Bergen; 12 February 2020

[17] Making CO_2 Storage a Viable Business for Oil and Gas Companies. CO_2 Storage Capture and Business Model. Petromall; July 2019

[18] JPT 27/08/2020 HSE Now Article Oxy Forms New Company to Pull Carbon Dioxide from Air. 2020

Section 3

Surfactants Flooding

Chapter 5

An Overview of Natural Surfactant Application for Enhanced Oil Recovery

Afeez Gbadamosi, Adeyinka Yusuff, Augustine Agi and Jeffrey Oseh

Abstract

Surfactant flooding is an enhanced oil recovery (EOR) method that recovers residual and capillary trapped oil by improving pore scale displacement efficiency. Due to toxicity and high cost of conventional surfactant, recent trend involves the use of natural surfactant for EOR. Natural surfactants are benign and biodegradable as they are derived from plant leaves and oil extracts. Herein, a synopsis of recent trend in the incorporation of newly devised natural surfactant for EOR was reviewed. Experimental results show that the surfactants exhibited sterling properties desired for EOR such as lower adsorption, interfacial tension (IFT) reduction, stable emulsion, and wettability alteration of sandstone and carbonate rocks. Overall, natural surfactants are suitable replacement for conventional surfactant. Nonetheless, an accurate modeling and pilot scale studies of natural surfactants remain obscure in literature.

Keywords: surfactant, natural surfactant, biosurfactant, enhanced oil recovery, wettability, interfacial tension

1. Introduction

Global demand for oil and gas continues to increase despite the recent development in other sources of energy. The production of oil and gas is in stages. Firstly, hydrocarbons are produced from reservoir deposit due to pressure reduction in the reservoir. Thereafter, waterflooding is performed to push more oil towards the production well. Substantial amount of oil is left in the reservoir as bypassed and/or residual oil after the application of primary and secondary recovery techniques. This is adduced to viscous fingering phenomenon as the injected waterflood creates a path of least resistance to the production well. Hence, several enhanced oil recovery (EOR) methods have been devised to recover additional oil from the reservoir [1]. **Figure 1** depicts the classification of EOR processes.

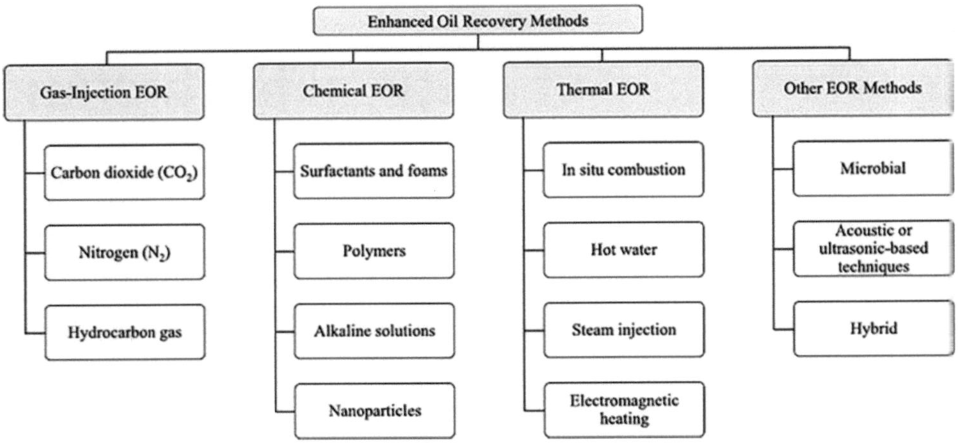

Figure 1.
Enhanced oil recovery process classification [2].

The EOR methods are broadly categorized into thermal and non-thermal EOR methods [3]. Thermal EOR are majorly used for the recovery of heavy oil, extra-heavy oil, and tar sands in the reservoir. Several thermal injections have been explored and exploited to improve recovery of high viscosity oils. These include cyclic steam stimulation, steam flooding, steam-assisted gravity drainage, and *in-situ* combustion. The mechanism of thermal EOR is to use high temperature to reduce the high viscosity and consequently improve the mobility of the oil towards the production well. Despite the success recorded for the field application of thermal EOR techniques during field application, they are deemed unsuitable for reservoir with huge depth and thin pay zones. Besides, they have high energy consumption, and large CO_2 emissions, thereby, increasing the environmental and economic costs of application [4]. Thus, non-thermal EOR methods have recently received prodigious attention.

The non-thermal EOR methods are gas EOR, microbial EOR and chemical EOR [5, 6]. Gas EOR methods involves the injection of miscible, immiscible, or inert gases into the reservoir to improve recovery factor. In addition to improving recovery factor, the use of gas injection also aids sequestration of gas in subsurface geologic formations. The mechanism of gas flooding includes the mass transfer of components between the oil and injected gas, swelling and viscosity reduction. The application of gas flooding is limited for high viscosity oils because of gravity override. Microbial EOR entails the use of microorganisms and their metabolites to mobilize capillary trapped oil. This method of EOR is cost-effective as it utilizes cheap raw materials from corn syrups, molasses, and agricultural by-products. Unfortunately, the raw materials for microbial EOR require huge facilities for their cultivation and have limited application due to high sensitivity and logistic problems especially on offshore platform [7].

Chemical EOR methods are adjudged to have a high efficiency, thus, they have witnessed numerous field applications [8]. The method basically involves tuning the efficiency of the injected waterflood to alter the rock-fluid and/or fluid-fluid properties of the reservoir. Hence, a high pore displacement and/or sweep efficiency is achieved, and consequently a higher recovery factor. The chemicals injected for EOR include alkaline, surfactant, polymer and more recently nanoparticles [9, 10]. Sometimes, a binary combination of the chemicals may be used to explore the synergic mechanism of the chemicals for a higher oil recovery. Several binary combinations used comprises of alkaline-surfactant flood, alkaline-polymer flooding, surfactant-polymer flooding, and alkaline-surfactant-polymer flooding [11–14]. The mechanism of alkaline and surfactant lowers the interfacial tension (IFT) between the oleic and aqueous phase and alter the wettability of the porous media, thereby, improving the pore scale displacement efficiency. On the other hand, the polymers thicken the viscosity of the injectant and, thus, reduce the mobility ratio and enhance the sweep efficiency [9, 10, 12, 15].

Surfactant flooding involves the use of surfactant molecules also referred to as surface active agents which are amphiphilic molecules with a hydrophilic (polar) head and hydrophobic (non-polar) tail [16, 17]. The hydrophobic tail, which is mostly oil soluble, is characterized by a long chain of alkyl groups which may be branched. The hydrophilic head which are mostly water soluble consists of moieties and are classified according to their ionic charge. Surfactant aid microscopic displacement efficiency by reducing IFT of the fluid-fluid interface. By reducing the IFT, the capillary forces of the trapped oil are minimized, and the oil saturation decreases [18]. Consequently, the dimensionless capillary number increases, and the recovery factor increases. Additionally, wettability alteration at the rock-fluid interface, foam generation and emulsification at the oil-interface are other mechanisms through which surfactant aid oil recovery. Surfactants alter wettability of the porous media via coating and/or cleaning mechanism. Besides, the emulsion generated ensure conformance control by creating a stable front while foam generated diverts subsequently injected water to thief zones in the reservoir to aid additional oil recovery [19].

Surfactants are classified based on the hydrophilic head group. The conventional surfactants based on ionic charge are the cationic surfactant, anionic surfactant, zwitterionic (amphoteric) surfactant, and non-ionic surfactant [20]. For cationic and anionic surfactants, the hydrophilic head groups are positively and negatively charged, respectively. Non-ionic surfactants have no charge and, hence, do not ionize in solution but are soluble in water due to the presence of hydrogen bonding between the hydrophilic groups [18]. Zwitterionic surfactant consists of hydrophilic head with positive and negative charges. Recently, the design and use of new surfactant such as Gemini surfactant, viscoelastic surfactant and polymeric surfactant have been exploited for EOR. Gemini surfactant is a surfactant composed of two single-chain surfactants linked by a spacer. The properties of the Gemini surfactants are dictated by the type and length of the spacer [21]. On the other hand, viscoelastic and polymeric surfactants are surfactants that form a supramolecular structure in solution and characterized by a high viscosity with additional ability to decrease interfacial tension which are both beneficial for EOR [22]. Nonetheless, the major issues associated with the use of conventional surfactants are high costs and environmental concerns. With recent issues associated with global warming and persistent regulation to lower environmental impacts on climate change, the industry is dissipating more energy and drive towards surfactant chemicals with less toxicity.

More recently, natural surfactants have received prodigious attention due to their lower toxicity, biodegradable, and environmentally benign, and good efficiency at improving recovery efficiency. Herein, natural surfactant is defined as surfactant synthesized from plants and oil. Natural surfactants possess the property of reducing the surface and interfacial tension in similitude to synthetic surfactants. Additionally, the novel surfactant has shown exemplary characteristics of foaming, emulsification, dispersion and wetting which are desirable for EOR. Herein, a synopsis of the application of natural surfactant and biosurfactant application for EOR was elucidated.

2. Natural surfactant

Natural surfactants are surfactants synthesized from leaves of plants and oils and they have shown sterling properties for use in EOR. Natural surfactants are either extracted directly or they are synthesized from plants and animal fats. Several methods have been reported for the synthesis of this biodegradable surfactants such as spray drying, freeze drying, Soxhlet extraction process, supercritical CO_2 extraction, ultrasonic extraction, microwave extraction, hydrolysis, and esterification process [23]. Moreover, several parts of plants such as leaves, roots, seeds, oils, and flowers have been courted for natural surfactants depending on their constituent components. *Jatropha curcas*, almond seed, *Vernonia amygdalina*, *Ziziphus spina-christi*, palm tree, vitagnus plant, and soapnut plant are several plants that have been exploited for natural surfactants [24–28]. Additionally, oils of plants and animals such as palm oil, coconut oil, rapeseed oil, sesame oil, waste cooking oil, linseed oil, and waste chicken fats have been converted into natural surfactants via esterification process [29–32]. Amino acid is another source of natural surfactant and can be derived from both plants and animals [23]. In similitude to conventional surfactants, the synthesized and/or extracted surfactants can be categorized into non-ionic, anionic, cationic, and zwitterionic surfactants. Other categories of synthesized polymeric, Gemini and viscoelastic surfactants [33, 34].

3. Application of natural surfactant for EOR

3.1 Interfacial Tension (IFT)

Interfacial tension is the adhesive tensional force that exists between the molecules of oil and water in porous media that ensures they remain trapped in the pores of the reservoir rock system. To improve recovery factor, the capillary force holding the oil in place must be minimized. This is achieved by lowering the IFT which in turn cause an increase of the capillary number and resultantly cause the residual oil to flow towards the oil bank and later to the production well. When the surfactant is injected into the reservoir rock system, due to their amphiphilic nature, the hydrophilic head of the surfactant aligns with the water and/or brine while the hydrophobic tail attaches with the oleic phase. The IFT of oil-water interface is measured in the laboratory via pendant drop or spinning drop method. Natural surfactant has demonstrated suitability for use in reducing IFT of oil-water interface. An important property of natural surfactant which makes it highly applicable for IFT reduction is its good solubility.

Several studies have demonstrated the viability of natural surfactant for lowering the IFT of crude oil-water interface. Yekeen et al. [26] investigated the IFT and foaming property of natural surfactant extracted from *Sapindus mukorossi*. At high temperature and pressure (80 °C and 8 MPa), the natural surfactant reduced the IFT of the oil-water interface from 23.24 mN/m to 1.59 mN/m. Moreover, the foam stabilized by the saponin-based natural surfactant was stable and perform comparatively well to conventional sodium dodecyl sulfate SDS-stabilized foam. Bahraminejad et al. [35] examined the IFT and foaming characteristic of surfactant extracted from *Gundelia tournefortii* plant. The surfactant lowered the IFT from 28 mN/m to 3 mN/m and generated stable foams. Imuetinyan et al. [36] evaluated the performance of natural surfactant extracted from *Vernonia Amygdalina* at oil-water interface. The natural surfactant lowered the IFT at oil-brine interface from 18 mN/m to 0.97 mN/m in the presence of NaCl brine. Additionally, the emulsion stabilized by the natural surfactant remained stable for longer periods. This implies that the use of the natural surfactant as injectant will ensure good conformance control in the reservoir.

Additionally, natural surfactant has demonstrated good stability and sterling properties at high salinity and high temperature conditions. Zhang et al. [37] synthesized a natural zwitterionic surfactant from castor oil and evaluated the salt tolerance and thermal stability and foaming performance. At 33,000 mg/l salinity and 50 °C, the synthesized surfactant maintained strong interfacial activity, which demonstrates good use for EOR. Zhang et al. [32] studied the performance evaluation of novel bio-based surfactant from waste cooking oil. The synthesized bio-based zwitterionic surfactant reduced oil-water IFT to 0.0016 mN/m at low dosage of 0.1 g/l. Kumar et al. [38] noted that bio-based polymeric surfactant synthesized from Jatropha oil was stable in 2.5 wt.% brine concentration and reduced the IFT of the oil-water interface from 22.4 mN/m to 0.33 mN/m. Hence, the application of bio-based natural surfactant shows good potential for EOR due to their stability and sterling interfacial properties at harsh condition typical of reservoir condition.

3.2 Wettability

Wettability is a pore scale displacement property and usually defined as the tendency of a fluid to spread on or adhere to a surface in the presence of other immiscible fluid. In similitude to conventional surfactant, natural surfactant has shown good property for altering the wettability of porous media from oil-wetting condition to water-wetting condition. Water-wetting condition is desired for better recovery efficiency (see **Figure 2**). Using contact angle measurement, Deng et al. [39] defined water-wetting condition as contact angle of 0 °–70 °, intermediate wetting condition as 70 °–110 °, and oil-wetting condition as contact angle 110 °–180 °. By modifying the wettability of the rock substrate to water-wetting condition, the capillary adhesive force that strongly attaches the oil to the rock diminishes, thus, allowing oil to flow. The interaction of the surfactant and the rock may cause alteration of the wettability condition depending on the type of surfactant injected and the porous media.

Numerous studies have indicated the ability of natural surfactant to cause wettability alteration of porous media. Imuetinyan et al. [25] observed that the natural surfactant extracted from *Vernonia Amygdalina* altered the wettability of sandstone surface from 118.5 ° to 45.7 °. Similarly, Singh et al. [40] evaluated the wettability of surfactant extracted from Fenugreek seeds. The surfactant reduced the IFT of oil-water interface to 44 °. Zhang et al. [37] reported wettability alteration from 92.04–38.79 ° using natural surfactant synthesized from castor oil. Moreover, natural surfactant synthesized from soybean oil reduced the wettability of rock substrate by 52.08% to 44.1 ° [41].

Figure 2.
Wettability alteration of reservoir rock system [9].

Chen et al. [34] synthesized a thermally stable and salt tolerant natural surfactant from waste cooking oil and evaluated it wettability alteration potential. The biobased surfactant altered the wettability of simulated formation water containing 0.5 g/l of the surfactant from 96.17 °–30.7 °. The contact angle decreased further to a meager 27.8 ° on further increment of the surfactant concentration to 3.0 g/l.

Furthermore, Kumar et al. [24] synthesized natural surfactant from Jatropha oil and evaluated the wettability on oil-wet quartz surface. The synthesized surfactant altered the wettability of the quartz surface to water-wetting condition. Despite the ability of natural surfactant to alter wettability to water-wetting condition, many of the natural surfactant do not alter the wettability to strongly water-wetting condition (contact angle <30 °). Hence, future research should focus on modifying the structure of the natural surfactant to improve it interaction and efficiency with reservoir rock system. Moreover, wettability studies of natural surfactant on porous media focused more on their behavior on quartz surface and sandstone. More studies of natural surfactant behavior on carbonate (calcite, dolomite, and limestone) surface are required.

3.3 Adsorption

Adsorption is an important property that demonstrates the economic viability of the chemical flooding process [42]. Low retention of chemicals is desired to ensure an economic and cost-effective recovery process. Due to their amphiphilic nature, the injection of surfactant into porous media is followed by the interaction of the surfactant with the rock via electrostatic, van der Waals, ion exchange and association, polarization of the π electrons, and hydrophobic interaction depending on the type of surfactant. The adsorption of natural surfactants has been studied on sandstone and carbonate reservoir rocks and showed minimal adsorption desired for chemical EOR process [43]. Yusuf et al. [44] studied the adsorption behavior of natural surfactant from soapnut fruit on carbonate rocks by batch adsorption experiments using surface tension techniques. They reported lower retention of the natural surfactant compared to ionic surfactants (cationic CTAB and anionic SDS). This was attributed to the weaker hydrogen bonding of the non-ionic surfactant.

Additionally, Kesarwani et al. [45] examined the adsorption property of novel biodegradable surfactant synthesized from Karanj oil on sandstone. The anionic surfactant had 15% lower retention compared to conventional anionic surfactant (SDS). Abbas et al. [46] performed a comparative study of saponin-based natural surfactants from Fenugreek, Sugar beet leaves and chickenpeas on quartz sand surface using UV-Vis spectrophotometer. They reported lower adsorption of the natural surfactant in high salinity conditions and adduced this phenomenon to the compact configuration of the surfactant and the compression of the electrical double layer. Ahmadi et al. [47] investigated the utilization of natural surfactant extracted from *Ziziphus spina-christi* (ZSC). Adsorption studies of the surfactant on carbonate rock samples via batch tests. The surfactant showed no sign of precipitation, but the presence of salt cations increases the adsorption of the surfactant due to electrostatic attraction force between the positive charge of the carbonate rock surface and the negative charge of the hydroxyl group on the surfactant.

3.4 Oil displacement

The major aim of deploying surfactant as a chemical injectant is to boost oil recovery. Oil displacement test is used in the laboratory to estimate recovery factor of injectant. Several studies have been carried out to ascertain the oil recovery potential

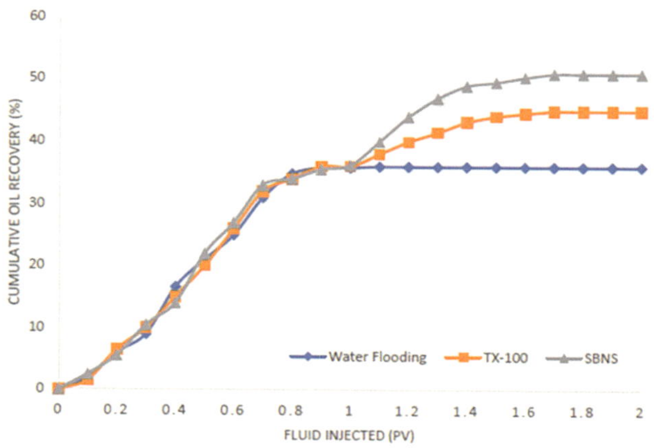

Figure 3.
Cumulative oil recovery of surfactant flooding [36].

of natural surfactant when used as injectants in sandstone and carbonate rocks. Saxena et al. [48] evaluated the oil recovery potential of surfactant synthesized from palm oil. The injection of 0.5 pore volume (PV) of surfactant in sandpack caused an improved recovery factor of 25–27% over conventional waterflooding. Nowrouzi et al. [49] evaluated natural surfactant synthesized from *Myrtus communis*. The surfactant yielded 14.3% incremental oil recovery when injected in carbonate core plugs. Alsabagh et al. [30] investigated oil displacement properties of green surfactant synthesized from waste cooking oils. 0.4 wt.% of the surfactant generated from palm kernel oil yielded 24.3% incremental oil recovery, respectively. Imuetinyan et al. [36] recorded 15% incremental oil recovery from core flooding procedure of saponin-based natural surfactant performed at high-pressure high-temperature condition. The SBNS performed better than conventional Triton X-100 under the same condition as depicted in **Figure 3**. Ahmadi et al. [50] explored the use of surfactant derived from *Glycyrrhiza glabra* for EOR. 8 wt.% of the newly extracted surfactant yielded 36% incremental oil recovery. Nafisifar et al. [29] applied surfactant synthesized from linseeds in sandstone cores. The natural surfactant injected after water flooding process yielded a 7.9% incremental oil recovery. Notwithstanding the numerous research of the application of natural surfactant for oil recovery, some aspects still need to be clarified in subsequent research. Notably, the concentration required for some natural surfactants are extremely high which may make the EOR process uneconomical.

4. Conclusion

This paper reviews the previous studies on natural surfactant application for EOR. The natural surfactants are benign and biodegradable and offer an alternative for existing conventional surfactants. Experimental studies show that natural surfactant can lower the IFT at the oil water interface. Moreover, the application of natural surfactant alters wettability of oil-wet cores to water-wetting condition. Adsorption studies of natural surfactant show that natural surfactant exhibit moderate retention behavior in reservoir cores and compares well to existing conventional surfactants. Moreover, oil displacement studies confirm that the application of natural surfactant can improve the recovery factor. Future studies on natural surfactant should focus on modeling their flow and transport behavior in porous media and scaling up natural surfactant application for field studies.

Conflict of interest

The authors declare no conflict of interest.

Author details

Afeez Gbadamosi[1*], Adeyinka Yusuff[2], Augustine Agi[2] and Jeffrey Oseh[3,4]

1 College of Petroleum and Geosciences, King Fahd University of Petroleum and Minerals, Dhahran, Saudi Arabia

2 Department of Chemical and Petroleum Engineering, Afe Babalola University, Ado-Ekiti, Nigeria

3 Department of Petroleum Engineering, Universiti Teknologi Malaysia, Johor, Malaysia

4 Department of Petroleum Engineering, Federal University of Technology Owerri, Nigeria

*Address all correspondence to: aogbadamosi2@live.utm.my

References

[1] Kamal MS, Adewunmi AA, Sultan AS. Recent advances in nanoparticles enhanced oil recovery : Rheology , interfacial tension , oil recovery , and wettability alteration. Journal of Nanomaterials. 2017;**2017**:15

[2] Massarweh O, Abushaikha AS. The use of surfactants in enhanced oil recovery: A review of recent advances. Energy Reports. 2020;**6**:3150-3178

[3] Agi A, Junin R, Gbadamosi A. Mechanism governing nanoparticle flow behaviour in porous media: Insight for enhanced oil recovery applications. International Nano Letters. 2018;**8**:1-29

[4] Guo K, Li H, Yu Z. In-situ heavy and extra-heavy oil recovery: A review. Fuel. 2016;**185**:886-902

[5] Gbadamosi AO, Kiwalabye J, Junin R, Augustine A. A review of gas enhanced oil recovery schemes used in the North Sea. Journal of Petroleum Exploration and Production Technologies. 2018:1-15

[6] Gbadamosi A, Radzuan J, Manan M, Agi A, Oseh J. Nanotechnology application in chemical enhanced oil recovery. In: Current Opinion and Recent Advances. London, UK: IntechOpen; 2019. pp. 26-50

[7] Niu J, Liu Q, Lv J, Peng B. Review on microbial enhanced oil recovery: Mechanisms, modeling and field trials. Journal of Petroleum Science and Engineering. 2020;**192**:107350

[8] Gbadamosi AO, Junin R, Manan MA, Agi A, Yusuff AS. An overview of chemical enhanced oil recovery: Recent advances and prospects. International Nano Letters. 2019;**9**(3):171-202

[9] Gbadamosi AO, Junin R, Manan MA, Yekeen N, Agi A, Oseh JO. Recent advances and prospects in polymeric nanofluids application for enhanced oil recovery. Journal of Industrial and Engineering Chemistry. 2018;**66**:1-19

[10] Gbadamosi AO, Junin R, Manan MA, Yekeen N, Augustine A. Hybrid suspension of polymer and nanoparticles for enhanced oil recovery. Polymer Bulletin. 2019;**76**(12):6193-6230

[11] Bashir A, Sharifi Haddad A, Rafati R. A review of fluid displacement mechanisms in surfactant-based chemical enhanced oil recovery processes: Analyses of key influencing factors. Petroleum Science. 2021;**19**(3):1211-1235

[12] Agi A, Junin R, Gbadamosi A, Manan M, Jaafar MZ, Abdullah MO, et al. Comparing natural and synthetic polymeric nanofluids in a mid-permeability sandstone reservoir condition. Journal of Molecular Liquids. 2020;**317**:113947

[13] Fu L, Zhang G, Ge J, Liao K, Pei H, Jiang P, et al. Study on organic alkali-surfactant-polymer flooding for enhanced ordinary heavy oil recovery. Colloids and Surfaces A: Physicochemical and Engineering Aspects. 2016;**508**:230-239

[14] Gao C, Shi J, Zhao F. Successful polymer flooding and surfactant-polymer flooding projects at Shengli Oilfield from 1992 to 2012. Journal of Petroleum Exploration and Production Technologies. 2014;**4**:1-8

[15] Gbadamosi AO, Junin R, Abdalla Y, Agi A, Oseh JO. Experimental investigation of the effects of silica

nanoparticle on hole cleaning efficiency of water-based drilling mud. Journal of Petroleum Science and Engineering. 2019;**172**:1226-1234

[16] Elhag HH, Abbas AH, Gbadamosi A, Agi A, Oseh J, Gbonhinbor J. Evaluation of continuous surfactant flooding in North East Africa: Case Study of Bentiu Reservoir Niger. Annu. Int. Conf. Exhib. Orig. Sched. Victoria Island, Lagos, Niger, 2020p. 1-11.

[17] Abbas AH, Sulaiman WRW, Jaafar MZ, Gbadamosi AO, Ebrahimi SS, Elrufai A. Numerical study for continuous surfactant flooding considering adsorption in heterogeneous reservoir. Journal of King Saudi University. 2018;**32**(1):91-99

[18] Kamal MS, Hussein IA, Sultan AS. Review on surfactant flooding: Phase behavior, retention, IFT, and field applications. Energy & Fuels. 2017;**31**:7701-7720

[19] Sagbana PI, Abushaikha AS. A comprehensive review of the chemical-based conformance control methods in oil reservoirs. Journal of Petroleum Exploration and Production Technologies. 2021;**11**:2233-2257

[20] Pal S, Mushtaq M, Banat F, Al Sumaiti AM. Review of surfactant-assisted chemical enhanced oil recovery for carbonate reservoirs: Challenges and future perspectives. Petroleum Science. 2018;**15**:77-102

[21] Kamal MS. A review of gemini surfactants: Potential application in enhanced oil recovery. Journal of Surfactants and Detergents. 2016;**19**:223-236

[22] Afolabi F, Mahmood SM, Yekeen N, Akbari S, Sharifigaliuk H. Polymeric surfactants for enhanced oil recovery: A review of recent progress. Journal of Petroleum Science and Engineering. 2022;**208**:109358

[23] Atta DY, Negash BM, Yekeen N, Habte AD. A state-of-the-art review on the application of natural surfactants in enhanced oil recovery. Journal of Molecular Liquids. 2021;**321**:1-24

[24] Kumar S, Kumar A, Mandal A. Characterizations of surfactant synthesized from Jatropha oil and its application in enhanced oil recovery. AICHE Journal. 2017;**63**:2731-2741

[25] Imuetinyan H, Agi A, Gbadamosi A, Junin R, Oseh J. Oil-water interfacial tension, wettability alteration and foaming studies of natural surfactant extracted from Vernonia Amygdalina. Petroleum Research. 2021. In press

[26] Yekeen N, Malik AA, Idris AK, Reepei NI, Ganie K. Foaming properties, wettability alteration and interfacial tension reduction by saponin extracted from soapnut (Sapindus Mukorossi) at room and reservoir conditions. Journal of Petroleum Science and Engineering. 2020;**195**:107591

[27] Emadi S, Shadizadeh SR, Manshad AK, Rahimi AM, Nowrouzi I, Mohammadi AH. Effect of using Zyziphus Spina Christi or Cedr Extract (CE) as a natural surfactant on oil mobility control by foam flooding. Journal of Molecular Liquids. 2019;**293**:111573

[28] Dashtaki SRM, Ali JA, Manshad AK, Nowrouzi I, Mohammadi AH, Keshavarz A. Experimental investigation of the effect of Vitagnus plant extract on enhanced oil recovery process using interfacial tension (IFT) reduction and wettability alteration mechanisms. Journal of Petroleum Exploration and Production Technologies. 2020;**10**:2895-2905

[29] Nafisifar A, Khaksar Manshad A, Reza SS. Evaluation of a new green synthesized surfactant from linseeds – chemical EOR implications from sandstone petroleum reservoirs. Journal of Molecular Liquids. 2021;**342**:1-22

[30] Alsabagh AM, Aboulrous AA, Abdelhamid MM, Mahmoud T, Haddad AS, Rafati R. Improvement of heavy oil recovery by nonionic surfactant/alcohol flooding in light of the alkane carbon number and interfacial tension properties. ACS Omega. 2021;**6**:18668-18683

[31] Pal N, Saxena N, Divya Laxmi KV, Mandal A. Interfacial behaviour, wettability alteration and emulsification characteristics of a novel surfactant: Implications for enhanced oil recovery. Chemical Engineering Science. 2018;**187**:200-212

[32] Zhang Q-Q, Cai B-X, Xu W-J, Gang H-Z, Liu J-F, Yang S-Z, et al. The rebirth of waste cooking oil to novel bio-based surfactants. Scientific Reports. 2015;**5**:1-7

[33] Babu K, Pal N, Saxena VK, Mandal A. Synthesis and characterization of a new polymeric surfactant for chemical enhanced oil recovery. Korean Journal of Chemical Engineering. 2016;**33**:711-719

[34] Chen Z-Z, Gang H-Z, Liu J-F, Mu B-Z, Yang S-Z. A thermal-stable and salt-tolerant biobased zwitterionic surfactant with ultralow interfacial tension between crude oil and formation brine. Journal of Petroleum Science and Engineering. 2019;**181**:1-7

[35] Bahraminejad H, Manshad AK, Keshavarz A. Characterization, micellization behavior, and performance of a novel surfactant derived from *Gundelia tournefortii* plant during chemical enhanced oil recovery. Energy & Fuels. 2021;**35**:1259-1272

[36] Imuetinyan H, Agi A, Gbadamosi A, Junin R. Extraction, characterization and evaluation of saponin-based natural surfactant for enhanced oil recovery. Arabian Journal of Geosciences. 2022;**15**:1-14

[37] Zhang Q-Q, Cai B-X, Xu W-J, Gang H-Z, Liu J-F, Yang S-Z, et al. Novel zwitterionic surfactant derived from castor oil and its performance evaluation for oil recovery. Colloids and Surfaces A: Physicochemical and Engineering Aspects. 2015;**483**:87-95

[38] Kumar S, Saxena N, Mandal A. Synthesis and evaluation of physicochemical properties of anionic polymeric surfactant derived from Jatropha oil for application in enhanced oil recovery. Journal of Industrial and Engineering Chemistry. 2016;**43**:106-116

[39] Deng X, Shahzad Kamal M, Patil S, Muhammad Shakil Hussain S, Zhou X, Mahmoud M. Wettability alteration of locally synthesized cationic gemini surfactants on carbonate rock. Journal of Molecular Liquids. 2021;**344**:1-13

[40] Singh A, Chaturvedi KR, Sharma T. Natural surfactant for sustainable carbon utilization in cleaner production of fossil fuels: Extraction, characterization and application studies. Journal of Environmental Chemical Engineering. 2021;**9**:1-15

[41] Xu W, Bian P, Gang H, Liu J, Mu B, Yang S. A novel bio-based sulfonic zwitterionic surfactant derived from transgenic soybean oil and its performance in surface and interfacial activities. Journal of Petroleum Science and Technology. 2018;**8**:32-44

[42] Tumba J, Agi A, Gbadamosi A, Junin R, Abbas A, Rajaei K, et al. Lignin as a potential additive for minimizing surfactant adsorption on clay minerals in

different electrolyte concentration. SPE-198713-MS, Pap Present SPE Niger Annu Int Conf Exhib Lagos, Niger, 2019:1-23.

[43] Kalam S, Abu-Khamsin SA, Kamal MS, Patil S. A review on surfactant retention on rocks: Mechanisms, measurements, and influencing factors. Fuel. 2021;**293**:120459

[44] Yusuf M, Wathon MH, Thanasaksukthawee V, Saul A, Tangparitkul S. Adsorption of Saponin natural surfactant on carbonate rock and comparison to synthetic surfactants: An enhanced oil recovery prospective. Energy & Fuels. 2021;**35**:11193-11202

[45] Kesarwani H, Saxena A, Saxena N, Sharma S. Oil mobilization potential of a novel anionic Karanj oil surfactant: Interfacial, wetting characteristic, adsorption, and oil recovery studies. Energy & Fuels. 2021;**35**:10597-10610

[46] Abbas AH, Abd Alsaheb RA, Abdullah JK. Comparative study of natural chemical for enhanced oil recovery: Focus on extraction and adsorption at quartz sand surface. Petroleum. 2022. In presss

[47] Ahmadi MA, Shadizadeh SR. Spotlight on the new natural surfactant flooding in carbonate rock samples in low salinity condition. Scientific Reports. 2018;**8**:1-15

[48] Saxena N, Pal N, Dey S, Mandal A. Characterizations of surfactant synthesized from palm oil and its application in enhanced oil recovery. Journal of the Taiwan Institute of Chemical Engineers. 2017;**81**:343-355

[49] Nowrouzi I, Mohammadi AH, Manshad AK. Preliminary evaluation of a natural surfactant extracted from Myrtus communis plant for enhancing oil recovery from carbonate oil reservoirs. Journal of Petroleum Exploration and Production Technologies. 2021;**12**:783-792

[50] Ahmadi MA, Galedarzadeh M, Shadizadeh SR. Wettability alteration in carbonate rocks by implementing new derived natural surfactant: Enhanced oil recovery applications. Transport in Porous Media. 2015;**106**:645-667

Chapter 6

Improving the Heavy Oil Recovery by Surfactants from Wastes

Ahmed Mohamed Al Sabagh and Asmaa Mohamed

Abstract

The amount of crude oil available must be sufficient to meet global demand. As a result, the oil industry has been obliged to recover oil from more difficult places and develop methods for enhanced oil recovery (EOR). This chapter focuses on the phase behavior properties inside the reservoir in connection with surfactant flooding and oil/brine systems in relation to enhanced oil recovery. To achieve this purpose, three groups of nonionic and anionic surfactants were prepared from waste and local materials. The surface activity and thermodynamic properties for three surfactant groups have been investigated at reservoir conditions. The solubilization parameters and relative phase volume were also studied to determine the optimal solubilization parameters and optimal salinity. The dynamic IFT and contact angle were measured at the optimal salinity. The sand pack flooding by using surfactant system predicted the performance of microemulsion in oil recovery by surfactant individually and its blends on chemical flooding system in semipilot EOR unit.

Keywords: green surfactants, waste materials, surface tension, interfacial tension, thermodynamic, adsorption mechanism, reservoir conditions, phase behavior, solubilization parameters, chemical flooding

1. Introduction

The amount of crude oil available must be sufficient to meet global demand. As a result, the oil industry has been obliged to recover oil from more difficult places and develop methods for enhanced oil recovery (EOR). This chapter focuses on the phase behavior properties inside the reservoir in connection with surfactant flooding and oil/brine systems in relation to enhanced oil recovery. To achieve this purpose three groups of nonionic and anionic surfactants were prepared from waste and local materials. The surface activity and thermodynamic properties for three surfactant groups have been investigated at reservoir conditions. The solubilization parameters and relative phase volume were also studied to determine the optimal solubilization parameters and optimal salinity. The dynamic IFT and contact angle were measured at the optimal salinity. The sand pack flooding by using surfactant system predicted the performance of microemulsion in oil recovery by surfactant individually and its blend and this applied on chemical flooding system in semi pilot EOR unit.

2. Enhanced oil recovery

There are several types of oil recovery mechanisms, including primary, secondary, and tertiary recovery. Primary recovery produces less than 20% of the original oil present in place (OOIP). In secondary recovery, the oil produced by water or gas flooding to make pressure maintenance. The final stage of oil production is tertiary recovery, also known as enhanced oil recovery (EOR). Then the oil recovery falls into the following three categories:

- **Primary recovery**: Recovery by depletion (natural pressure of reservoirs)
- **Secondary recovery**: Recovery by injecting water or gas flooding
- **Tertiary recovery:** Recovery of the residual oil by injecting additives not present in reservoir

Improved oil recovery (IOR): Improved oil recovery (IOR) is a broad term that refers to a variety of activities. Improved oil recovery (IOR), which is synonymous with EOR, refers to any process or practice that improves oil recovery. IOR includes EOR processes, also include other practices, such as waterflooding, pressure maintenance, infill drilling, and horizontal wells.

Enhanced oil recovery(EOR): Injection of fluids that are not present in reservoirs such as surfactants. EOR is the process of injecting external materials into reservoirs to manage interfacial tension (IFT), fluid characteristics, wettability, and overcome pressure retention forces in order to recover trapped crude oil from pores and transport it to a production well. The capacity to control the flow of displacement fluids, also known as mobility control, is one feature of EOR operation that has significant effects on all processes. In order to commercialize the EOR process, its economic viability is more vital than any other factor [1, 2].

2.1 EOR processes

Water flooding is the most widely used technique and has been used for a long time. The water flooding does not remove all of the oil from the production zone. The tertiary recovery then becomes the main goal for producing residual oil. The oil that remains after primary and secondary recovery is distributed throughout the reservoir's pores. Oil trapping is primarily caused by capillary and viscous forces.

The common classifications of different EOR processes are:

- Chemical EOR
- Thermal EOR
- Miscible/Immiscible EOR
- Microbial EOR
- Technical Challenges and Futures Techniques in EOR

2.2 Chemical EOR or chemical flooding

Chemical processes involve the injection of a specific chemical liquid that effectively creates desirable phase behavior properties in order to improve oil displacement. These displacing fluids have low interfacial tension (IFT) with the displaced crude oil. The main chemicals used in EOR chemical flooding are alkaline, surfactants, and polymers. Each material has a certain mechanism for enhancing the oil flow properties. Alkali behaves as in situ surfactants, where the alkali function groups react with the naphthenic carboxylic groups of crude oil forming in situ sodium salt surfactants. The surfactants are prepared on the surface and then injected inside reservoirs. They improve the oil production by reducing the IFT between crude oil and connate water forming an emulsion that has low viscosity and the ability to make wettability alteration. In case of polymers, they added to the displacing water to increase its viscosity in order to control and make sweeping for the residual oil present in porous media, therefore, the oil production efficiency increases. Generally, there are several types of polymers that are used in this field. The most commercially attractive polymers are polyacrylamides (PAM) and polysaccharides (Biopolymers). In chemical EOR flooding process, hydrolyzed polyacrylamides (HPAM) give higher viscoelasticity, and they are preferred over polysaccharides [3, 4].

2.3 Surfactant flooding

One of the most promising methods for increasing oil productivity from low-pressure reservoirs is surfactant-assisted enhanced oil recovery. Surfactant flooding is an approved EOR technique for getting residual oil out of a reservoir. The goal of surfactant injection into the reservoir to improve the oil recovery factor is to change the fluid/fluid interaction by lowering the IFT between oil and brine, as well as the fluid/rock properties by changing the wettability of the porous medium, or a combination of both mechanisms. The hydrophilic head dissolves with water when surfactant solutions are poured into oil reservoirs with brine, whereas the hydrophobic tail reacts with crude oil components. The adsorbed film is produced as a result of the interaction between the oil and the alkyl tail of the surfactant, thus lowering the IFT at the oil/water interface. The mechanism by which a surfactant alters the wettability of conventional rock pores is known as a cleaning mechanism, in which the surfactant adsorbs at the oil-wet layer and then changes the surface wettability from oil-wet to water-wet. In addition to having a high surface activity and wettability, good surfactants should be biodegradable and nontoxic. The surface activity and thermodynamic properties provide information on the arrangement of surfactant molecules between two phases and the reduction of surface tension. The micellization in bulk and adsorption at interface can be studied by Gibb's isotherm. Micellization and adsorption are important in understanding the factors that affect CMCs values, such as structural effect. When surfactant concentrations reach critical micelle concentrations (CMCs), micelles develop. Reservoir parameters like pressure, temperature, and salinity of formation water influence CMC and interfacial phenomena (IFT). IFT is one of the most measured parameters to be lowered to less than 10^{-2}mN / m [5–7].

2.3.1 Surfactant flooding mechanism

Surfactant flooding improves pore-scale displacement efficiency through the mechanism of interfacial tension reduction, wettability alteration, or a combination of both mechanisms.

2.3.1.1 Interfacial tension reduction

Due to oil entrapment by capillary forces, it is nearly impossible for water to displace all of the oil in the pore scale during secondary recovery by water flooding. This capillary force is measured by a dimensionless capillary number (Nc) defined in Eq. (1) as:

$$Nc = \mu\nu/\sigma \cos\Theta \quad (1)$$

where **μ** is the displacing fluid viscosity, **v** is the displacing Darcy velocity, **Θ** is the contact angle, and **σ** is the **IFT** between the displacing fluid (water) and the displaced fluid (oil). **Nc** is closely related to residual oil saturation and oil recovery and increases as residual oil saturation decreases. Consequently, a higher **Nc** will result in a higher oil recovery A typical brine flooding has a **Nc** in the range of 10^{-7} to 10^{-6}. From Eq. (1), this can be achieved in three ways: (1) increasing the displacing fluid viscosity (**μ**); (2) increasing the injection fluid velocity (**v**); (3) reducing the IFT (**σ**).

2.3.1.2 Wettability alteration

Wettability is the tendency of a solid surface to attract a specific type of fluid in the presence of other immiscible fluids. The position, distribution, and movement of fluids inside a reservoir rock system are determined by the wettability of the rock surface. Most oil reservoirs are classified as oil-wet, water-wet, or mixed wet. Surface imaging tests, zeta potential measurements, spontaneous imbibition, and contact angle measurements can all be used to assess this feature of the reservoir rock system. The contact angle, which is defined as the point where the interface of the oil and water meets at the rock surface, is used in the majority of studies of wettability alteration measurements. A surface with a contact angle greater than 90° is considered oil-wet, while a surface with a contact angle less than 90° is considered water-wet. Changing the wettability of a surface from oil-wet to water-wet reduces capillary adhesive force and increases reservoir oil permeability [8, 9].

2.3.2 Surfactant types and their structure

EOR has investigated a number of surfactants for use in oil recovery. They are classified into anionic surfactants, nonionic surfactants, cationic surfactants, and zwitterionic surfactants depending on the nature of the hydrophilic head group.

2.3.2.1 Anionic surfactant

Anionic surfactants are the most commonly used surfactants. The majority of surfactant flooding EOR work has been done on sandstone reservoirs. The surface-active portion of this class of surfactant bears a negative charge such as carboxylate

(COO^-), sulfate (SO_4^-), or sulfonate (SO_3^-), though in association with a cation usually an alkaline metal (Na^+ or K^+).

2.3.2.2 Cationic surfactant

Cationic surfactants are surfactants that have a positive charge on their hydrophilic head, but only in conjunction with a halide group. In water, they split into an amphiphilic cation and an anion. This surfactant class is easily attracted to the negatively charged surfaces of rocks and is very effective at changing reservoir rock wettability [10].

2.3.2.3 Nonionic surfactant

Nonionic surfactants, unlike cationic and anionic surfactants, do not ionize in aqueous solution. Alcohol, phenol, ether, ester, and amide are examples of non-dissociable hydrophilic functional groups. Meanwhile, the lipophilic group consists of the alkyl or alkyl benzene group. Despite the lack of ionic charge, the hydrophilic group is soluble in water due to its inherent polarity induced by the presence of hydrogen bonds and van der Waals interactions. Nonionic surfactants have a higher salinity tolerance than ionic surfactants; however, they have a lesser IFT reduction.

2.3.2.4 Zwitterionic surfactant

Zwitterionic surfactants are characterized by the presence of anionic and cationic surface charges on their hydrophilic head. When they dissociate, they display anionic and cationic characteristics. They can also withstand high salinity and high temperatures. Betaine and sulfobetaine are two examples of this type of surfactant.

3. Microemulsions

3.1 Type and structure of microemulsion

The structure of the microemulsion plays an important role in the physicochemical properties of the applied fields. Direct (oil dispersed in water, o/w), reversed (water dispersed in oil, w/o), and bi-continuous microemulsions are the three basic types. Multiple microemulsion, like multiple emulsion, is sometimes possible. The structure of a microemulsion is determined by salinity, water content, co-surfactant concentration, and surfactant concentration [11].

3.2 Applications in enhanced oil recovery (EOR)

In oil and gas industry, the approach to emulsion and/or microemulsion preparation has been associated with the application of energy to a mixture of oil, water, and emulsifier. Because of the rheological and thermodynamical properties of emulsions, injection of emulsions and/or microemulsions into oil reservoirs has been recognized as a potential tool for oil recovery [12].

3.3 Microemulsions for enhanced oil recovery

Microemulsions could also be used to improve oil recovery because of the ultra-low interfacial tension values achieved between the contacting oil and water microphases. Microemulsion flooding can be applied over a wide range of reservoir conditions. Microemulsion techniques involve pumping water into the oil reservoir that contains a small amount of surfactant and other chemicals. The natural acids in the trapped oil react with this solution to form a microemulsion. The surfactant molecules break down the interfacial tension to mobilize oil and enable it to escape from the rock. Microemulsions are prepared from a mixture of oil, water, or brine and a surfactant. In some cases, the addition of a co-surfactant (alcohol) is required to ensure the stability of the microemulsion. An oil-in-water (O/W) microemulsion in equilibrium with the oil excess phase (Winsor I), a water-in-oil (W/O) microemulsion in equilibrium with the water excess phase (Winsor II), and a microemulsion in equilibrium with both the water and oil excess phases (Winsor III) are prepared for a given overall composition. Surfactant flooding operations are best performed with middle-phase microemulsions. Hence, it is fundamental to maintain the middle microemulsion phase as long as possible during the process of surfactant flooding. Many factors influence the best surfactant composition for a microemulsion system, including pH, salinity, temperature, and so on. Due to the ability to dissolve oil and water concurrently, as well as the system's ability to achieve very low interfacial tension, tertiary oil recovery using microemulsions has been the main focus. Microemulsion flooding is a miscible displacement procedure that optimizes oil recovery by reducing capillary forces on oil droplets in the reservoir [13–15].

3.4 Surfactant flooding: Optimum phase type and optimum salinity

As salinity rises, surfactants are able to solubilize an increasing amount of oil and a diminishing amount of water. The salinity at which the microemulsion solubilizes equal amounts of oil and water is called the optimal salinity. Salinity scan tests are commonly used to assess the phase behavior of surfactant formulations before conducting time-consuming core-flood testing. When the minimum interfacial tension is linked to the solubilization parameters at the optimal salinity, the presence of viscous, structured phases, and stable macroemulsions can be easily monitored. The equilibrium phase behavior appears to shift from a lower-phase microemulsion to an upper-phase microemulsion over a narrow salinity range. Depending on salinity, a microemulsion can exist in three types of systems: type I, type III, or type II. The system is type I below a certain salinity. The system is classified as type II above a certain salinity. If the salinity is in between, the system is type III. The interfacial tension (IFT) of microemulsion/brine is lower in a type III system than in a type I system, and the IFT of microemulsion/oil is lower in a type II system. At optimum salinity, the two IFTs are equal. If the optimum salinity decreases with surfactant concentration, it will also decrease as the surfactant solution progresses. As a result, as the surfactant solution progresses, the decreasing salinity will be consistent with the decreasing optimum salinity, ensuring that the optimum salinity is maintained. Therefore, The oil recovery factor in a type III system is higher than in a type I or type II system. Core flooding must be used to establish the optimal phase type. The phase type with the highest oil recovery factor is the optimum salinity type [16, 17].

In this chapter two types of surfactants have been synthesized the first one is nonionic surfactants derived from polyethylene glycol having different molecular weights of 400,600,1000 and 2000 with either mixed fatty acids of jatropha oil and waste cooking oil or dodecylbenzene sulfonic acid and the other is anionic surfactants, which are derived from either mixed fatty acid of the oil used or dodecylbenzene sulfonic acid. The chemical structure confirmation of the prepared surfactants was recorded using a Thermo Fisher Scientific Spectrometer (400–4000 cm^{-1} Nicolet Is^{-10}) (FT-IR spectra). The phase behavior and solubilization parameters of the prepared surfactants were studied. The phase behavior of surfactant-brine-oil system in the oil recovery by microemulsion system was evaluated. Finally, sets of flooding experiments for the prepared surfactants and their blends with and without co-surfactants (Iso Propanol) on sand-packed model at critical micelle concentration (CMC), different temperature and different salinities was performed.

4. Materials

- Purified jatropha oil with a density of 0.908 g/cc and an acid value of 11.87 KOH/gm was used, as well as waste cooking oil with a density of 0.989 g/cc and an acid value of 16.73 KOH/gm was obtained from an Egyptian market. Para toluene sulfonic acid (99%), anhydrous sodium sulfate (99%), and HCl (99%) were obtained from Sigma-Aldrich Company. The solvents, ethanol (99%), n-Hexane (99%), isopropanol (99%), and xylene (99%) were purchased from PIOCHEM Company. The polyethylene glycols (400, 600, 1000, and 2000) were obtained from Alfa Aesar Company. Alkyl benzene sulfonic acid, KOH, and NaCl were purchased from El-Gomhouria Co, Egypt.

- The crude oil used in microemulsion preparation was obtained from General Petroleum Company, Egypt. API gravity at 60 °F 27, asphaltine content 7.63 wt. %, saturates 57wt%, resin content 15 wt. %, aromatic 21wt%, and density is 0.8983 at 15°C.

- The TDS of formation water was 200×10^3 ppm. Also, formation water was diluted by distilled water to get a salinity of 50×10^3 and 100×10^3 ppm. The 50×10^3 and 100×10^3 ppm salinity is prepared to represent the formation water's salinity and study the effect of changing salinity on prepared microemulsions.

As the results of EOR operations, this work interested to prepare three groups of surfactants to be used in this application. The abbreviation of these groups was; group1 (EABS9, 14, 23, 46 and EABSNa), group2(EHJ9, 14, 23, 46 and EHJNa), group3 (EHWO9, 14, 23, 46 and EHWONa). All of these research work results, data analyses and comparative studies are elaborated on and discussed in this chapter [18].

5. Surface active properties of surfactants prepared at 25°C

The surface tension of the three surfactant groups was measured in the formation water at a temperature of 25 ° C. The first group of surfactants was based on dodecyl benzene sulfonic acid (**G1**), the second group on jatropha oil (**G2**), and the third group on waste cooking oil (**G3**). The breakpoint of the plots was used to determine the values of (CMC) and surface tension at CMC (γ_{CMC}). G1 has lower CMC values than the other two groups, which could be related to the presence of a sulfonic group in the molecules, which deactivates the surfactant molecules' solubility in the solution. The values of CMC in relation to the number of ethylene oxides in the tested groups were found to decrease with an increase in the number of ethylene oxide units until a certain number was reached, after which they increased. This behavior may be caused by two factors. The first factor is the ethylene oxide chain coiling, which influences the solution's solubility. The second factor could be due to surfactant molecule solubility in formation water as a result of salts in the water breaking down hydrogen bonds. These results of CMC show that the G2 and G3 can efficiently saturate any interface, demonstrating surface properties like flexibility and low interfacial tension that can be used in EOR. The surface tension was reduced by the surfactant molecules, allowing for a quantitative investigation that revealed continual adsorption at the interface. So that, at concentrations lower than the CMC, the possibility of micelle production is not fully realized. The effectiveness (π_{CMC}) is the difference between the surface tension values of the formation water only and with surfactant at CMC and determined by Eq. (2):

$$\pi_{CMC} = \gamma_w - \gamma_s \tag{2}$$

where γ_w is the surface tension of formation water and γ_s is the surface tension of surfactant solution at CMC. By calculating the average of π_{CMC}, it was found that the π_{CMC} of G1 was higher than the other two groups. The maximum surface excess concentration (Γ_{max}) is the maximum amount up to which surfactant adsorption can be obtained at the surface, and this depends on the molecular structures of the interacting component. The adsorption degree was calculated by Gibb's isotherm and given by Eq. (3):

$$\Gamma_{max} = -(1/RT)\,(\delta\gamma/\delta \ln C) \tag{3}$$

where $\mathbf{\Gamma_{max}}$ is the surface excess concentration (mol/cm^2), T is the absolute temperature, **R** is a universal gas constant (8.314 Jmol^{-1}K^{-1}), and ($\boldsymbol{\delta\gamma/\delta \ln c}$) is the slope of γ-lnC. The minimum area occupied by surfactant molecules ($\mathbf{A_{min}}$) determines the average area occupied by each adsorbed surfactant molecule at the air-liquid interface at saturated surface. The $\mathbf{A_{min}}$ was determined by Eq. (4):

$$A_{min} = 1 \times 10^{16}/(NA.\Gamma_{max}) \tag{4}$$

where **NA** is the Avogadro's number (6.022×10^{23}). The results of A_{min} are decreased by increasing the ethylene oxide units for the three groups, but the A_{min} values of G1 are lower than those of G2 and G3. This could be because the surfactant molecules in the G1 have three chemical spaces; the benzene ring, the SO_2, and the ethylene oxide chain, this may result in vertical adsorption of molecules and the formation of a monolayer on the surface or interface, lowering the coiling affinity of the ethylene oxide chain and lowering the A_{min} by increasing the units of ethylene oxide. The coiling affinity of the ethylene oxide chain in these groups may be obtained by two factors. The first factor is the formation of water which inhibits the formation of hydrogen bonds, lowering the solubility of surfactant molecules, and causing coiling. The second explanation could be related to the unsaturation of the double bond in the oleic chain, which results in cis and trans configurations, increasing the ethylene oxide chain's coiling affinity. The low values of A_{min} show that the ability for the formation of oil/surfactant/solution microemulsion resulting in lowering the interfacial tension, further the oil displacing capacity should be improved. The adsorption efficiency (Pc_{20}) is given by the negative logarithm of the surfactant concentration that reduces the surface tension of the pure solvent by 20mN/m. The adsorption efficiency is determined by Eq. (5):

$$Pc_{20} = -\log C_{20} \qquad (5)$$

where C_{20} is the amount of surfactant required for reduction of pure solvent surface tension by 20 mN/m and this means that C_{20} is the minimum concentration that denotes the adsorption saturation at the surface. Therefore, C_{20} measures the efficiency of surfactant molecules' adsorption at the air-liquid interface. The higher the Pc_{20} number, the more surfactant molecules adsorb. The surfactants of G1 achieved the lowest concentration in terms of adsorption efficiency. The surface pressure was in the maximum value with G3. This means that surfactants can successfully saturate any interface by displaying surface properties with the appropriate flexibility while also lowering the IFT.

6. Surface active properties of EABS14, EHJ23, and EHWO14 at different temperatures

Three surfactants were chosen, one from each group, to demonstrate how temperature affects their surface and thermodynamic properties. The selectivity of these surfactants is determined by how much surface tension and area per molecule are reduced (A_{min}). When the temperature was raised, the CMC of these surfactants increased slightly, while the surface tension decreased. This could be due to warmth breaking the hydrogen bond, making the surfactant molecules more soluble in the solvent, resulting in higher concentration consumption and adsorption. Temperature increased adsorption efficiency, indicating that the surfactants' high surface activity was responsible. The effectiveness decreased with an increase in temperature. When the temperature was raised, the A_{min} exhibited a small increase. The decrease in Γ_{max}

Surfactant	**EABS14**				**EHJ23**				**EHWO14**			
Temperature	**25°C**	**35°C**	**45°C**	**55°C**	**25°C**	**35°C**	**45°C**	**55°C**	**25°C**	**35°C**	**45°C**	**55°C**
CMC × 10^{-3} (mol/L)	1.38	1.58	1.69	1.80	1.98	1.98	2.17	2.38	2.89	3.02	3.12	3.22
Pc_{20} (mol/L) × 10^{-4}	0.454	3.35	4.44	5.53	0.167	1.67	3.33	5.00	0.432	4.49	8.39	12.3
γ_{CMC} (mNm^{-1})	25.9	22.3	20.35	18.4	26.5	22.5	43.4	18.9	28.5	25.5	22.45	19.4
Γ_{max} ($mol/cm^2 \times 10^{-10}$)	2.49	1.73	1.51	1.29	1.84	1.38	1.28	1.18	1.89	1.54	1.50	1.46
A_{min} ($nm^2 \times 100$)	0.665	0.96	1.11	1.278	0.900	1.199	1.29	1.400	0.879	1.08	1.11	1.14
π_{CMC} ($mN\ m^{-1}$)	41.53	33.41	29.69	25.98	40.93	29.21	27.36	25.51	38.93	30.21	27.62	25.02

Table 1.
Surface activity for EABS14, EHJ23, and EHWO14 in formation water at different temperatures [18].

values and increase in A_{min} could be owing to thermal agitation caused by repulsive forces between bulk phase molecules. The repulsive forces in the bulk phase are based on the breakdown of hydrogen bonds (**Table 1**).

7. Thermodynamic properties of the prepared surfactants

Surface tension measurements were used to calculate the micellization and adsorption free energy at the interfaces. The adsorption of surfactant molecules at the air-liquid interface under equilibrium conditions reduces surface tension. The number of surfactant molecules adsorbed at the interface per unit area was provided by Gibbs adsorption equation. The CMC values play a vital role in calculating ΔG_{mic}. This is shown in the following Eq. (6):

$$\Delta G_{mic} = RT \ln_{CMC} \quad (6)$$

where ΔG_{mic} is the molar Gibbs energy of micellization in KJ/mol. The change in the adsorption free energy was calculated from Eq. (7):

$$\Delta G_{ads} = \Delta G_{mic} - [0.6022 \times \Pi_{cmc} \times A_{min}] \quad (7)$$

The production of micelles in the bulk phase of the solution was indicated by negative ΔG_{mic} values. Negative ΔG_{mic} values indicate that micellization is a spontaneous association dissociation process that allows surfactant molecules adsorbed at the interface. At the same time, the negative values of ΔG_{mic} increase the free energy of the solvent, which compensates for the surfactant molecules prefer to adsorb on the surface and interface before and during micelle formation. The negative values of ΔG_{ads} indicated that the adsorption of the surfactant molecules at the air-liquid interface is a spontaneous process. Due to an increase in the curvature of the air/aqueous surface, the greater negative values of ΔG_{ads} increase. This means that as temperature rises, the number of potential vacancy sites for adsorption increases, and more surfactant molecules should be adsorbed at the interface. Surfactant molecules establish a microemulsion at the CMC for all surfactants tested, and as a result of negative ΔG_{mic} and ΔG_{ads} values, the surfactant molecules may form a stable microemulsion phase and demonstrate effective interfacial contact with the surrounding media. Based on the micelle aggregation and adsorption capabilities of these three surfactant groups, it is expected that they will contribute to the right formulations to generate microemulsions for use in oil solubilization and displacement processes in the enhanced oil recovery field (**Tables 2** and **3**).

Surfactant	**Δ G_{mic} (KJ mol^{-1})**				**Δ H_{mic} (KJ mol^{-1})**				**Δ S_{mic} (KJ^{-1} mol^{-1}k^{-1})**
	25°C	**35°C**	**45°C**	**55°C**	**25°C**	**35°C**	**45°C**	**55°C**	
EABS14	−16.32	−16.51	−16.74	−16.97	−7.38	−7.27	−7.27	−7.28	0.03
EHJ23	−15.43	−15.94	−16.08	−16.22	−6.49	−6.70	−6.61	−6.53	0.03
EHWO14	−14.48	−14.86	−15.13	−15.41	−2.56	−2.54	−5.03	−2.49	0.04

Table 2.
Thermodynamic parameters of micellization for EABS14, EHJ23, and EHWO14 in formation water at different temperatures [18].

Surfactant	ΔG_{ad} (KJ mol^{-1})				ΔH_{ad} (KJ mol^{-1})				ΔS_{ad} (KJ^{-1} mol^{-1}k^{-1})	ΔG_{mic} -ΔG_{ads} (KJ mol^{-1})			
	25°C	35°C	45°C	55°C	25°C	35°C	45°C	55°C		25°C	35°C	45°C	55°C
EABS14	−31.50	−35.82	−36.39	−36.96	28.10	25.78	26.71	27.64	0.21	15.81	19.31	19.72	19.99
EHJ23	−37.61	−37.03	−37.38	−37.73	−34.63	−33.95	−34.22	−34.5	0.01	22.18	21.09	21.16	21.51
EHWO23	−35.09	−34.45	−33.52	−32.58	−5.29	−3.65	−1.96	−0.28	0.1	20.61	19.59	18.39	17.17

Table 3.
Thermodynamic parameters of adsorption and structural effects of EABS14, EHJ23, and EHWO14 in formation water at different temperatures [18].

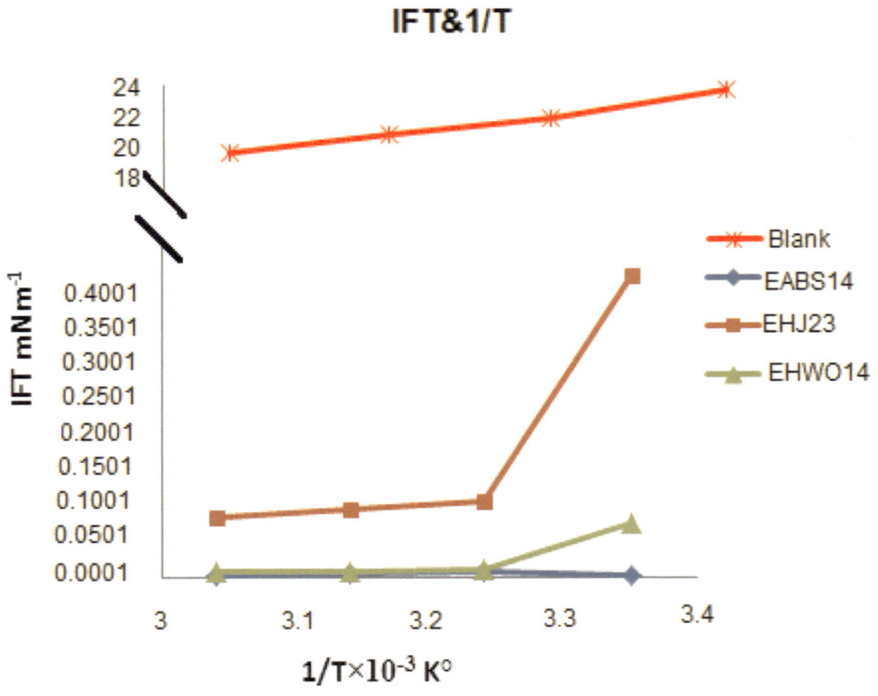

Figure 1.
IFT for Blank, EABS14, EHJ23, and EHWO14 at Different Temperatures [18].

8. Interfacial tension in the reservoir condition

When assessing surfactant effectiveness for oil recovery, the interfacial tension is a more important factor to consider. The lower the IFT value, the better the ability to generate microemulsion, which is more effective in displacing oil stuck in reservoir pore throats. The interfacial tension (IFT) was measured for the EABS14, EHJ23, and EHWO14 between the used formation water and crude oil by using Attension Theta High-Pressure Chamber (Sessile Method) (ASTM ISO 19403-5) to evaluate their affinity in the enhanced oil recovery application (EOR). The IFT was measured at different temperatures (25, 35, 45, and 50°C) in high saline formation water TDS (200×10^3 ppm) at the CMC. The data showed that by increasing temperature the IFT decreased. The data also ranged from 10^{-1} to 10^{-4}mN /m and these results were considered suitable for the application of these surfactants in the EOR. By increasing the temperature, the IFT decreased marginally [19]. This is because the temperature increases the free energy of the surfactant system, which helps to push the surfactant molecule to adsorb on the interface, resulting in oil solubilization in the form of microemulsion and, as a result, higher oil recovery (**Figure 1**).

9. Solubilization parameters and phase behavior

The volume of solubilized oil divided by the volume of solubilized surfactant in the microemulsion is the oil solubilization ratio. Similarly, the water solubilization ratio in a microemulsion is defined as the volume of water solubilized divided by the amount of solubilized surfactant. The difference in volume between the initial aqueous phase and excess water is used to calculate the volume of solubilized water. The optimum solubilization occurs when the solubilization of oil and water are equal. The solubilization curves are generated using data relevant to each tube. Water solubilization method with salinity variation can be used to determine the phase behavior and phase boundary of a microemulsion system. From the experimental results, it was found that when the salinity increases the solubilization values increase up to a certain value and then decreases. The salinity at which the solubilization is highest is termed optimal salinity. In the present study, the optimal salinity was found at 100×10^3ppm. At the optimal salinity, the middle phase of microemulsion has the ability to solubilize equal amounts of oil and brine. After the optimal salinity, the water solubilization decreases by increasing salt concentration. As the salinity increases, the microemulsion phase changes from Winsor type I to Winsor type II to Winsor type III. These phenomena can be illustrated on the basis of the interaction of the inter droplets and interfacial bending stress. As salt concentration increases, salt ions attract some water molecules, reducing the number of water molecules available to interact with the charged component of the surfactant and raising the demand for solvent molecules. As a result, the contact between the surfactant's hydrophilic head groups becomes stronger than in solution. Then the interfacial film turns from positive value to zero to negative value and this corresponds to phase transition from oil water (O/W) Winsor type I to bicontinuous phase Winsor type III to water oil (W/O) Winsor type II so increasing salinity causes phase transition from lower to middle to upper phase of microemulsion [20, 21].

10. Effect of salinity on the IFT

In case of oil/water system, the IFT was found to be high. The use of salt causes a significant shift in the IFT. The IFT between oil and microemulsion dropped as salt concentration was raised, however, the IFT between water and microemulsion increased. After certain concentration, the IFT microemulsion and oil increased. The minimum IFT

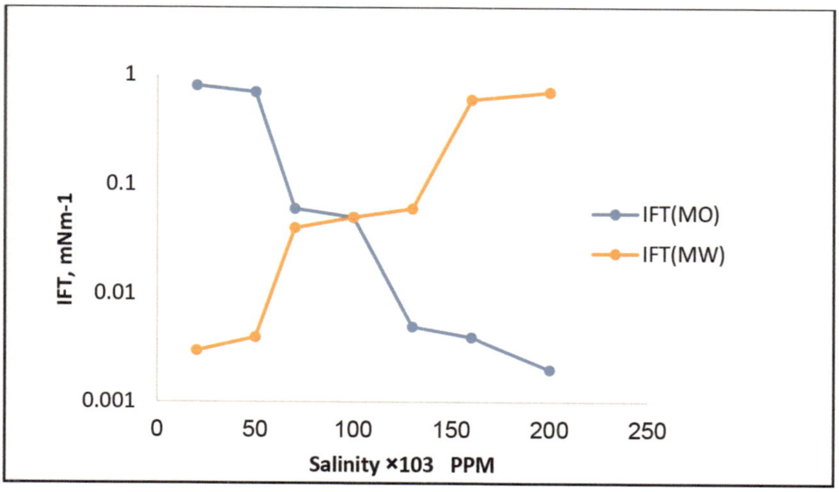

Figure 2.
Interfacial Tension Vs. Salinity for EHWO14+EABS14+Cs [18].

is obtained at certain salinity called optimal salinity. The reduction of IFT in presence of salt is interpreted as the following. Surfactant materials are responsible for lowering the IFT (interfacial tension) between oil and water. The presence of salt in the aqueous phase increases the concentration of surface-active species that are present in crude oil at the crude oil/water interface then lowering the IFT. Above the optimal salinity, the salts prevent the molecules of surfactant from dissolving in aqueous phase because of the increasing repulsive forces of electrostatic double layer [22]. As a result, the amount of surfactant in the oil phase was lowered, and the IFT could not be reduced (**Figure 2**).

11. Phase diagram of micro emulsion system

The pseudo ternary diagram of surfactant, co-surfactant/brine/crude oil system has been constructed for different types of surfactant in this study. The brine is considered as a single pseudo component, (S+Cs) is another single component, and crude oil is the last component. The importance of the construction of ternary diagram is to determine the composition of microemulsion. It is also important to prepare the microemulsion with low concentration of surfactant from economical point of view. The ternary diagrams for surfactant/ formation water/crude oil were constructed for surfactants and their blends. It was found that the microemulsion region in case of blend is larger than the area in surfactant only (**Figure 3**). This may be due to the presence of isopropanol co-surfactant, which increase the solubility of the surfactant molecules in the oil phase and increase the stability of microemulsion. In general, the microemulsion region of surfactant and its blends derived from waste cooking oil is larger than the microemulsion region of surfactant and its blends derived from jatropha oil. This may be due to by measuring the surface and IFT of surfactant solution it was found that the

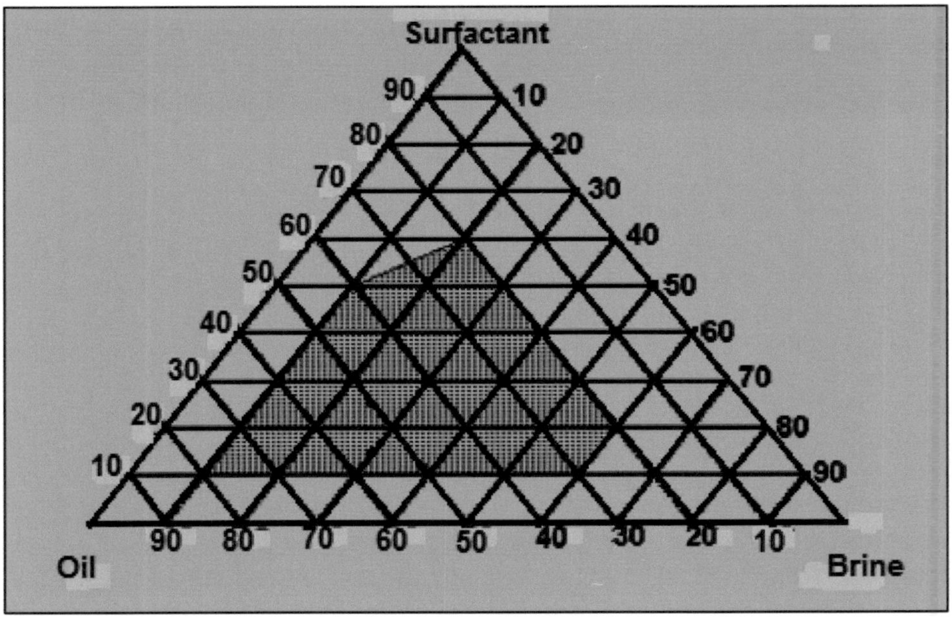

Figure 3.
Ternary Phase Diagram of Oil-Brine-Surfactant Blend (EHWO14+EABS14+Cs) System [18].

IFT values of surfactant derived from waste cooking oil are lower than the IFT values of surfactant derived from jatropha oil. This also may be due to the higher adsorption of surfactant molecules at the interface and this is shown from the thermodynamic properties where the ΔG_{ads} of surfactant derived from waste cooking oil group is more negative than ΔG_{ads} the surfactant derived from jatropha oil.

12. Enhanced oil recovery factor of the surfactant flooding

The successful surfactant flooding as a chemically EOR process is to design the surfactant slug at an optimum surfactant concentration. The flooding process depends on many factors, such as temperature, the critical micelle concentration (CMC), adsorption properties on the sand-packed model, interfacial tension, contact angle, and alteration wettability at the surface or the interface between core and the formation water. Different sets of flooding experiments for the EHJ23, EHWO14, and their blends with and without co-surfactants (isopropanol) were drawn with the injected pore volume of sand-packed model at CMC concentration, different temperatures (50, 70°C), and different salinities (50×10^3, 100×10^3, 200×10^3ppm). After the flooding of the surfactant solution, the trapped oil in the pores is mobilized due to the decrease in the IFT between the oil and the injecting surfactant solution. So that it can interact with the trapped oil and reduce the IFT and solubilize the oil by forming oil-in-water emulsion and changing the rock wettability, further the recovery factor (RF) increases. This indicates that the behavior of surfactant molecules causes complete adsorption and a stable electric double layer at the interface and that the IFT is minimized so that the maximal solubilization of oil by surfactant is obtained at the CMC, optimal salinity, and optimum temperature. The RF dropped after CMC, indicating that an increase in surfactant molecules led to the formation of a multilayer of surfactant adsorption, which could lead to the formation of an inverse emulsion, resulting in a reduction in the RF (**Figure 4**).

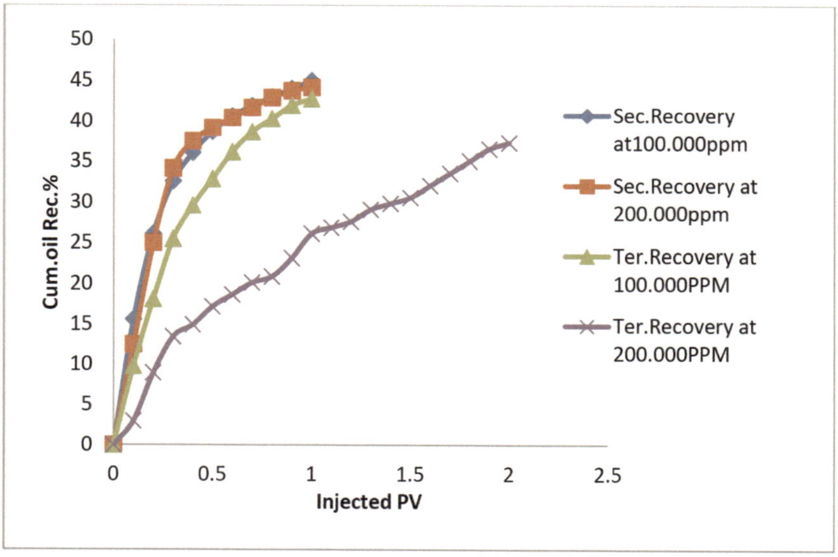

Figure 4.
Cumulative Oil Recovery Vs. Injected Pore Volume for EHWO14 at 70°C.

13. Conclusion

In this chapter, attention has been paid to prepare some anionic and nonionic surfactants from waste and nonedible materials to evaluate their performance in enhanced oil recovery (EOR). The surface activity, thermodynamic properties, and interfacial tension for surfactants have been investigated under reservoir conditions. The phase behavior of surfactant-brine-oil system is an important key in evaluating the oil recovery by microemulsion system so the phase behavior and solubilization parameters of the prepared surfactants were studied. The solubilization parameters for oil in microemulsion Vo/Vs are increasing as a function of salinity, whereas the solubilization parameters of water Vw/Vs are decreasing and this is shown for surfactants (EHWO14+EABS14+Cs). The optimal salinity was found to be 100×10^3 ppm and the minimum IFT is obtained at optimal salinity. The pseudo ternary diagram of surfactant, co-surfactant/brine/crude oil system has been constructed for surfactants (EHWO14+EABS14+Cs). Sets of flooding experiments for the EHWO14 and their blends were performed on sand-packed model at CMC concentration, different temperatures, and different salinities and maximum RF was achieved under these conditions.

Author details

Ahmed Mohamed Al Sabagh[1,2] and Asmaa Mohamed[1,2]*

1 Application Department, Egyptian Petroleum Research Institute, Cairo, Egypt

2 Enhanced Oil Recovery Unit (EOR) at EPRI, Egypt

*Address all correspondence to: ammnaser94@gmail.com

References

[1] Alvarado V, Manrique E. Enhanced oil recovery: An update review. Energies Journal. 2010;**3**:1529-1575

[2] Baojun B. EOR performance and modeling. Journal of Petroleum Technology (JPT). 2016;**68**:60-73

[3] Olajire AA. Review of ASP EOR (alkaline surfactant polymer enhanced oil recovery) technology in the petroleum industry: Prospects and challenges. Energy. 2014;**77**:963-982

[4] Thomas S. Enhanced oil recovery—an overview. Oil Gas Science and Technological Review. 2007; **63**:9-19

[5] Samanta A, Bera A, Ojha K, Mandal A. Comparative studies on enhanced oil recovery by alkali–surfactant and polymer flooding. Journal of Petroleum Explorer and Production Technology. 2012;**2**:67-74

[6] Bera A, Kumar T, Ojha K, Mandal A. Adsorption of surfactants on sand surface in enhanced oil recovery: Isotherms, kinetics and thermodynamic studies. Application Surface Science. 2013;**284**:87-99

[7] Panthi K, Weerasooriya U, Kishore K. Mohanty enhanced recovery of a viscous oil with a novel surfactant. Fuel. 2020; **282**:118-882

[8] Liu Q, Dong M, Ma S, Tu Y. Surfactant enhanced alkaline flooding for Western Canadian heavy oil recovery. A Physicochemical Engineering Aspects. 2007;**293**:63-71

[9] Chen W, David S. Schechter Surfactant selection for enhanced oil recovery based on surfactant molecular structure in unconventional liquid reservoirs. Journal of Petroleum Science and Engineering. 2021;**196**: 107702

[10] Pal S, Mushtaq M, Banat F, Al Sumaiti AM. Review of surfactant-assisted chemical enhanced oil recovery for carbonate reservoirs: Challenges and future perspectives. Petroleum Science. 2018;**15**:77-102

[11] Uma AB, Saaid IBM, AdebayoSulaimon A, Pilus RBM. A review of petroleum emulsions and recent progress on water in-crude oil emulsions stabilized by natural surfactants and solids. Journal of Petroleum Science and Engineering. 2018;**165**:673-690

[12] López-Montilla JC, Herrera-Morales PE, Pandey S, Shah DO. Spontaneous emulsification: Mechanisms, physicochemical aspects, modeling, and applications. Journal of Dispersion Science and Technology. 2002;**23**(1-3): 219-268

[13] Sharma T, Kumar GS, Sangwai JS. Enhanced oil recovery using oil-inwater (o/w) emulsion stabilized by nanoparticle, surfactant and polymer in the presence of NaCl. Geosystem Engineering. 2014;**17**(3):195-205

[14] Bera A, Mandal A. Microemulsions: A novel approach to enhanced oilrecovery: A review. Journal of Petroleum Exploration and Production Technology. 2015;**5**(3):255-268

[15] Li X, He G, Zheng W, Xiao G. Study on conductivity property and microstructure of Triton X-100/ alkanol/n-heptane/water microemulsion. Colloids and Surfaces A: Physicochemical and Engineering Aspects. 2010;**360**(1-3):150-158

[16] Santanna VC, Curbelo FDS, Castro Dantas TN, Dantas Neto AA, Albuquerque HS, Garnica AIC. Microemulsion flooding for enhanced oil recovery. Journal of Petroleum Science and Engineering. 2009;**66**(3, 4):117-120

[17] Zhao J et al. The structure effect on the surface and interfacial properties of zwitterionic sulfo betaine surfactants for enhanced oil recovery. The Royal Society of Chemistry. 2015;**5**:13993-14001

[18] Al Sabagh AM et al. Surface activity and thermodynamic properties of some green surfactants from wastes in formation water at reservoir conditions. Journal of Dispersion Science and Technology. 2020;**43**:385-398

[19] Al-Sabagh AM. Surface activity and thermodynamic properties of water-soluble poly ester surfactants based on 1, 3 dicarboxymethoxy-benzene used for enhanced oil recovery. Polymers for Advanced Technologies. 2000;**11**:48-56

[20] Bera A, Kumar T, Ojha K, Mandal A. Screening of microemulsion properties for application in enhanced oil recovery. Fuel. 2014;**121**:198-207

[21] Jafari S, Sofla D, Sharifi M, Hemmati A. Toward mechanistic understanding of natural surfactant flooding in enhanced oil recovery processes: The role of salinity, surfactant concentration and rock type. Journal of Molecular Liquids. 2016;**222**:632-639

[22] Roshanfekr M, Johns RT. Prediction of optimum salinity and solubilization ratio for microemulsion phase behavior with live crude at reservoir pressure. Fluid Phase Equilibria. 2011;**304**:52-60

Chapter 7

Chemical Enhanced Oil Recovery: Where Do We Stand? Recent Advances and Applications

Anastasia Ivanova and Alexey Cheremisin

Abstract

In this chapter, the recent advances in chemical flooding, including the application of nanoparticles, novel surfactants, and the combination thereof will be discussed and described. The main rock and reservoir fluids properties that influence the effectiveness of chemical flooding will be addressed. The emphasis will be given on wetting properties and recent advances in methods for measuring wettability. The technological and economic challenges associated with chemical injection will be posed, and recent solutions will be given. Especially, the challenge of applying chemical EOR methods to carbonate reservoirs will be covered, and suggestions to overcome it will be given. Moreover, the current worldwide applications of chemical EOR will be discussed and future plans will be outlined.

Keywords: chemical EOR, surfactants, polymers, nanoparticles, carbonate reservoirs, micro- and macro-wettability

1. Introduction

Alternative renewable energy sources, such as solar, wind, or hydrogen energy, are actively developing in the world. However, traditional oil and gas are still dominating sources of energy, and their global demand is growing continuously. Therefore, it is important to continue developing and enhancing recovery from existing oil fields or discovering new production fields [1]. However, the reserves of conventional, easily accessible hydrocarbons are consistently declining, which is attributed to the fact that the production of conventional oil has surpassed the increase of its reserves. Therefore, since the last several decades, there has been an increasing trend toward the development of unconventional sources of oil and gas, such as viscous (heavy) oils and bitumen, oil sands, oil and gas saturated low-permeability reservoirs, which will significantly contribute to reserves restocking. The majority of unconventional resources are deposited in remote regions with complex geological conditions (depth, porous media structure, mineralogical variations, etc.) under harsh reservoir properties, such as high temperature and salinity, and thus, their development involves the application of new technologies of exploration and recovery [2].

Several stages of recovery and reservoir development are known [3]. At the first stage, also called the primary stage, oil is extracted using the natural energy of the formation, due to which oil flows freely to the production well. However, over time, the initial reservoir pressure decreases, which consistently leads to a decrease in the oil recovery factor. In this case, secondary methods (or improved oil recovery methods) are used to maintain the pressure, such as the injection of water or gas into the reservoir. It is well known that after waterflooding more than 50% of residual oil will remain unproduced [4]. Such inefficient recovery after secondary methods is attributed to the reservoir's rocks and fluids properties, such as hydrophobic wetting properties and high oil viscosity. Therefore, enhanced recovery methods (or tertiary methods) are applied to change or modify reservoir properties of matured fields, facilitating the displacement of oil toward the production well. These methods include thermal formation stimulation (steam and air injection), chemical flooding (surfactant and polymer injection), gas injection (N_2 and CO_2), microbes injection, and combination of the methods. Generally, tertiary EOR technologies aid in incremental oil recovery more than primary and secondary methods [5]. However, the effective deployment of EOR methods requires a deep understanding of the mechanisms behind the fluid distribution and displacement through the porous media, which in turn control overall oil production.

In the face of diverse EOR advances, chemical methods are one of the promising techniques applied to recover residual and trapped oil [6, 7]. However, due to some challenges associated with chemical mismatch and high cost, porous trapping, and plugging issues, in the past decades, not many projects have been conducted in the fields. Nevertheless, the rising oil prices and market demand encourage many researchers around the world to further develop chemical EOR technology to make it more efficient yet cost-effective and environmentally friendly.

This review focuses on the fundamentals of chemical EOR in order to explain the main aspects behind screening procedures for chemicals and suitable reservoirs with an emphasis on surfactant injection. The work discusses the main types of studied surfactants and their properties (phase behavior, interfacial tension [IFT], and wettability) that should be evaluated prior to their application and the methods that are usually used. Subsequent sections describe the main advantages and challenges associated when surfactant flooding applied in sandstone or carbonate reservoirs. The novel approaches of using nanoparticles in surfactant flooding, so-called nanoEOR, will be addressed. Finally, examples of field applications will be given. This work mainly focuses on the overall properties of common chemicals and practical recommendations rather than on detailed descriptions of phenomena related to chemical EOR.

2. Fundamentals of chemical EOR

Various methods are being developed and applied to improve and enhance oil recovery from different reservoirs [2]. Conventionally, the effectiveness of the method or technology applied for reservoir development is assessed by means of the oil recovery factor (%) which is usually calculated by multiplying several factors:

$$\begin{aligned}\text{Recovery}(\%) = \text{pore} - \text{scale displacement}(\%) \times \text{sweep}(\%) \times \text{drainage}(\%)\\ \times \text{commercial cut} - \text{off}(\%),\end{aligned} \quad (1)$$

where pore-scale displacement is a measure of how much of the oil has been pushed out from any of the rock accessed by injected fluids, sweep calculates how much reservoir rock has been reached by injected fluids, drainage is an extent to which wells can access all the separate segments of the reservoir, and commercial cut-off indicates the limit of economic production.

Generally, pore-scale displacement and sweep efficiency are two main factors that can be controlled and modified by applying EOR methods. As a result, it facilities the oil displacement toward the production wells, increasing the overall recovery factor.

Waterflooding is the widely used method for decreasing residual oil saturation by pore-scale oil displacement. However, reservoirs show low oil recovery during waterflooding because more than half of the original oil in place (OOIP) is still left trapped in the complex pore matrix due to the low capillary number of water [8]. The capillary number is defined as the ratio between viscous and capillary forces (Eq. (2)) and controls the residual oil saturation [9] and as a result, the pore-scale displacement.

$$N_c = \frac{v * \mu}{\gamma * cos\theta}, \tag{2}$$

where μ is the brine viscosity, v is Darcy's velocity, θ is contact angle, and γ is IFT between oil and water phases.

It is well known [10] that to reduce residual oil saturation (i.e., enhance oil recovery), the capillary number should be increased drastically to 10^{-3} or higher from the typical number of waterflooding—10^{-7}. Indeed, in works [11, 12], it was shown that an increase in the capillary number to 10^{-4} to 10^{-3} reduces the residual oil saturation to 90%, and if the capillary number reaches about 10^{-2}, then the residual oil saturation tends to reach zero. Note that the relation between the capillary number and residual saturation is known as the capillary desaturation curve.

In practice, to modify the capillary number, chemical EOR methods, including surfactant flooding, are widely used. As it can be seen in Eq. (2), a capillary number can be increased in several ways: (1) by increasing the viscosity of the injected fluid; (2) by increasing injection fluid velocity; (3) by decreasing IFT between immiscible phases (water and oil); and (4) by decreasing a contact angle. However, an increase in the velocity of the injected fluid can lead to an undesirable increase in the injection pressure compared with the reservoir pressure. Therefore, more often, EOR methods are applied to change and modify the injected fluid viscosity, IFT, or wettability. For instance, the viscosity of injected water can be increased by adding long-chain molecules such as polymers, which due to the formation of a network of topological entanglements, impart high viscosity to the aqueous solution and in some cases, viscoelastic properties [13]. It is important to note that the main purpose of polymer addition is increasing sweep efficiency. Indeed, when a low-viscosity fluid (water) is injected into a reservoir, it will tend to bypass oil sections of the reservoir as it moves along, creating an uneven (fingered) profile. These fingers can have different shapes ranging from a "fleshy" finger [14] to a "skeletal" finger [15]. As a result, it will displace residual oil unevenly, leaving many pores with hydrocarbons untreated. Therefore, in order to reduce the mobility of the water and viscous fingering, polymers are added to the displacement fluid (water).

Furthermore, to reduce the IFT at the interface between two immiscible liquids, such as displacement fluid and oil, as well as to modify the wetting angle, surfactant solutions are used due to their unique properties [16]. Indeed, surfactant molecules are amphiphilic—they consist of hydrophobic (oil-soluble) and hydrophilic (water-soluble) parts. Due to amphiphilic properties, surfactant molecules can self-orient at the surface or interface via hydrophobic or electrostatic interactions, resulting in a reduction of surface energy. Depending on the charge of the hydrophilic group, surfactants are divided into ionic (charged) and nonionic (not charged). The most commonly used surfactants in chemical EOR are listed in **Table 1.** Conventionally, the type of surfactant for a specific reservoir is chosen accordingly to the screening procedure performed with reservoir rocks and fluids. A summary of general properties applicable to every type of surfactant is illustrated in **Table 1**.

Some types of surfactants have been found to be very effective in terms of IFT reduction, as such carboxylate surfactants can lower the IFT from 20 to 50 mN/m to 10^{-3} to 10^{-2} mN/m at reservoir conditions and thus increase the capillary number in 1000 times [17, 18]. The decrease in IFT between the displacement and displaced fluids makes the oil more mobile in the pore throats due to reduced capillary trapping.

Surfactant type by charge	Typical hydrophilic groups	Summary
Anionic (negatively charged)	Carboxyl ($RCOO^-X^+$), sulfonate ($RSO_3^-X^+$), sulfate ($ROSO_3^-X^+$), phosphate ($ROPO_3^-X^+$)	• The most well-studied type of surfactants • Low adsorption value on rocks with negatively charged surfaces (sandstone), but high on positively charged (carbonates) • Effective in IFT reduction and wettability alteration
Cationic (positively charged)	Ammonium, quaternary ammonium halides ($R_4N^+X^-$)	• Effective in wettability alteration • Less effective in IFT reduction than anionic • Low adsorption value on carbonate rocks • High cost
Non-ionic (no charge)	Polyoxyethylene, polyols, sucrose esters, polyglycidyl esters	• Usually used as a cosurfactant to improve the properties of ionic surfactants • Improve stability of ionic surfactants
Amphoteric (depending on the pH, can act differently in aqueous solutions—in an acidic solution they exhibit the properties of cationic surfactants, and in an alkaline solution—anionic surfactants)	Betaine, sulfobetaine, $RN^+(CH_3)_2CH_2CH_2SO_3^-$, imidazoline derivatives	• Effective in IFT reduction and wettability alteration • Have high stability at high temperatures and salinities • High cost

Table 1.
The list of typical surfactants used in chemical EOR and their general properties.

However, one of the main properties of surfactants that are lowering the IFT to values ~10^{-3} to 10^{-2} mN/m is significantly influenced by various factors, such as surfactant type and concentration, the concentration of monovalent and divalent ions in brine (salinity), water-oil ratio (WOR), oil composition, reservoir mineral composition, temperature, and pressure. Therefore, some surfactants that lower IFT at ambient conditions may not be applicable under reservoir conditions (high temperatures and mineralization).

Although many factors affect surfactants' ability of IFT reduction, their common property is the existence of optimal concentration, at which the minimum of IFT can be observed. This concentration is known as critical micelle concentration (CMC) and varies for different surfactants depending on their chemical structure. The CMC is the main controlling parameter for surfactant properties and needs to be optimized for every surfactant system for enhanced oil recovery [19]. Typically, this concentration lies in a range between 0.001% and 0.01% for commonly used surfactants. By increasing the surfactant concentration above the CMC point, the IFT curve starts to grow slightly [17, 19]. Notably, although, at the CMC point, the minimum IFT is reached, for practical cases concentration of surfactant is chosen of one–two orders of magnitude higher because of high retention and adsorption of surfactant molecules during penetration through the pore networks.

3. Wettability of formation rocks, its modification, and measuring methods

Capillary number can be also increased by reducing the contact angle $\cos\theta$ (Eq. (2)) that is determined by the forces between injected fluid and rock surfaces. This contact angle is defined in terms of rock-wetting properties. According to the general definition, wettability is the property of a liquid to maintain contact with a solid surface that stems from intermolecular interactions [20].

In terms of oil recovery, wettability is the main parameter that governs produced and injected fluids distributions in the porous networks, which in turn affects the properties of the multiphase flow of reservoir fluids. The wetting state of rock surfaces stems from a chemical equilibrium between rock/oil/brine that formed over years. Generally, wetting preferences of surfaces are divided into three types such as water-wet (hydrophilic), mixed-wet, and oil-wet (hydrophobic). In the water-wet state, oil forms droplet with the minimum contact area at the surface, showing a contact angle θ~0°. If oil drop spreads, the surface is considered hydrophobic or oil-wet with a contact angle close to 180°. When rock surfaces do not exhibit particular wetting preferences, the wetting state is referred to intermediate. In this case, the contact angle can be calculated by the balance of the surface tension forces between phases (Young equation) that determine the shape of a drop on the surface. Notably that when rock surfaces demonstrate several wetting preferences, for example, some areas are hydrophilic and others are hydrophobic, the wettability type is mixed.

The wetting state of reservoir rock controls the arrangement and migration of oil, brine, and gas throughout the pore channels. The distribution of wetting and nonwetting fluids depends on capillary forces, and thus, wetting fluid tends to occupy small pores while nonwetting remains in large pores. In terms of oil recovery, this means that if the rock exhibit hydrophobic wetting state, water as a non-wetting fluid will penetrate through the pores with a larger size, leaving small pores unaffected [18]. This phenomenon explains the low efficiency (low recovery factor) of waterflooding in hydrophobic oil reservoirs (i.e., carbonates), as the water moves through the big pores, while major oil resides in small pores, where water cannot access due to the capillary pressure effect.

This phenomenon is the main reason that prevents developing especially carbonate reservoirs by waterflooding, because the majority of them exhibit hydrophobic or oil-wet wetting preferences [21, 22]. Compared with sandstones, the wettability of carbonate reservoirs appears to be more complex. Indeed, the initially hydrophilic wetting state of carbonate rocks can be changed towards more hydrophobic when interactions between minerals and oil components take place. In research work, several possible mechanisms of wettability alteration due to such interactions have been proposed:

1. Adsorption of polar organic components or carboxylic acids from oil, such as asphaltenes, resin fractions, benzoic, palmitic, oleic, and octanoic acids [23–25];

2. Ion binding, which is led by the surface charge of carbonates [26].

It is important to estimate the initial wettability of rocks accurately for selecting a proper EOR technology for reservoir development. Conventionally, wettability studies are conducted at a core-scale by using Amott–Harvey [27], USBM [28, 29], and contact angle methods [30]. The popularity of these laboratory methods stems from their cost-effectiveness and simplicity. Although Amott–Harvey and USBM methods are widely used in laboratories, they provide information about the average wettability of the core sample. Furthermore, the investigation of reservoir rock wettability by these methods is limited, because as it was shown in work [31], only samples with permeability no less than 10 mD can be studied. Moreover, these methods cannot account for a mixed wettability state (case when different surfaces of rock exhibit both hydrophilic and hydrophobic wetting preferences), as it impossible to determine the number of areas, which have water-wet or oil-wet states, and thus they provide only the integral wettability index.

Another commonly used method for wettability investigation is measuring the contact angle between fluids (water or oil) and the surface. This method is based on the analysis of droplet shape when it spreads on the surface. In this case, wettability is assumed in terms of contact angle values. As such, wettability is referred to hydrophilic if oil forms an angle $0° < \theta < 70°$ with the surface, intermediate wet—$70° < \theta < 110°$, and hydrophobic—$110° < \theta < 180°$. The reverse trend of contact angle values is considered if using water instead of oil. The main difference between contact angle measurements and USBM and Amott–Harvey methods is that by measuring angles the data of wetting preference of a particular sample surface can be collected, and thus mixed wettability state can be determined correctly. However, the contact angle measurements cannot provide the average or integral wettability index. Therefore, in order to measure the wettability accurately, one should consider using a combination of different methods.

Nevertheless, direct study of the fluid-rock interactions by these methods is constrained, as they measure the average wettability on macroscale (mm) and cannot account for rock surface properties, such as its roughness, chemical composition, and pore structure that significantly influence fluids flow and distribution [32]. Therefore, in the past decades, advanced microscopic techniques, including high-resolution scanning electron microscopy (SEM), transmission electron microscopy (TEM) coupled with cryogenic technique and environmental scanning electron microscopy (ESEM) are proposed as new methods for the investigation of reservoir fluids arrangement in porous structure [33, 34]. For example, the authors [33] illustrated that by collecting the elemental maps of rock samples via coupled SEM imaging with the X-ray analysis of the elements the arrangement of reservoir fluids (brine and oil) in the porous network can be obtained. Furthermore, with the recent advances in microscopy, it has become possible to use cryogenic techniques (i.e., Cryo-TEM) and ESEM in order to study surface wettability at microscale (μm) [35, 36]. Indeed, in the work [37] authors carried out the wettability measurements of the middle Bakken samples using ESEM technique. The authors concluded that the results obtained at microscale could be applied for more accurate calculation of multiphase flow parameters (e.g., relative permeability and capillary pressure), which in turn would improve the development of primary or secondary oil recovery processes. Moreover, the application of microscopic techniques is particularly essential while developing the carbonate reservoirs, as these reservoirs show complex wetting behavior due to challenging pore structure, mineral composition, and heavy oil. Indeed studying the rock/brine/oil chemical interactions would give an insight into how to optimally modify wettability to mixed or water-wet, which in turn would increase oil recovery from carbonate reservoirs.

The authors [22] investigated the reason for the hydrophobic wetting properties of carbonate reservoir rocks using combined microscopic tools. The advanced microscopic technologies were first used to identify the adsorbed organic layers on rock surfaces that were proposed to be the key reason for hydrophobic wetting state of carbonates. It was shown [22] by using the ESEM approach that the surface had two wetting preferences. As such, the surface areas that were covered with hydrocarbon layers had hydrophobic wetting properties, while pure calcite areas exhibited hydrophilic state. This result was also confirmed by EDXS analysis of different areas. It was also revealed that the main parameters of multiphase fluids distribution in the pore channels, such as capillary pressure curves, could be evaluated more accurately when using data of microscale wettability variations and the thickness of the organic layers (180 ± 12 nm).

Furthermore, the obtained results can be used to explanation of the reasons for complex wettability behavior in carbonate reservoir rocks. Indeed, based on developed methods [22], it was suggested that some asphaltenes or oil acids could react with calcium ions on the surface by the ionic bond between calcium (Ca^{2+}) and oxygen (O^-) from the carboxyl group (COO^-). This is the initial layer of hydrophobic organic layers on carbonate surfaces, on which other oil hydrocarbons can adsorb, forming bigger hydrophobic regions. As a result, initial hydrophilic wettability alters toward more oil-wet and water-injection becomes ineffective. Importantly, this explanation can be also applied when developing other carbonate oil reservoirs with high content of carboxylic acids or asphaltenes in oil.

Furthermore, in [38], it was shown that the wettability of carbonates measured by a common laboratory method (contact angle) that differs from wettability measured in corresponding areas of the surface at the microlevel. As such, results demonstrated that wettability at the microlevel was mixed (i.e., hydrophilic and hydrophobic zones) while at the macrolevel surface showed only hydrophobic wetting preferences. These findings bring into question the applicability of macroscale data in reservoir modeling for enhanced oil recovery and geological storage of greenhouse gases.

Traditionally, wettability should be altered towards more hydrophilic in order to increase the oil recovery factor from hydrophobic reservoirs (including carbonates). As a result, spontaneous water imbibition into a porous media will be promoted, leading to the enhancement of oil recovery. In this regard, many different surface active chemicals, such as surfactants, have been widely tested for wettability alteration towards more water-wet. **Table 2** summarizes some literature data on change in water advancing contact angle after treating rock surfaces with various surfactants.

As can be observed from **Table 2**, the values of the contact angle between surface and water correspond to the hydrophilic wetting state after treatment of different surfactants.

Notably, the effectiveness of wettability alteration depends on molecular structure and the ionic nature of surfactants. For instance, it was shown that some anionic surfactants with ethoxy and proxy groups in a mixture with Na_2CO_3 were promising agents for alteration of wettability of carbonate surfaces from oil-wet to water-wet [42]. Contrary to this, the authors [39] suggested that cationic surfactants could be more effective for wettability changing in carbonates reservoirs than anionic ones. The authors explained this by the formation of ion pairs that occur between negatively charged oil components adsorbed on carbonate surfaces and positively charged surfactants. As a result, desorption of oil components from surfaces will be facilitated, leading to a consistent oil recovery increase. Furthermore, it was illustrated that wettability is altered more effectively due to the electrostatic interactions than by hydrophobic interactions. This hypothesis has been also supported by work [43], where the authors studied the wettability alteration process of carbonate cores using

Surfactant	Type	Contact angle, °		Ref.
		Before treatment	After treatment	
C12TAB	Cationic	70	12	[39]
C16TAB	Cationic		27	
C8TAB	Cationic		57	
C10TAB	Cationic		31	
Hyamine	Cationic		21	
Cropol	Anionic		55	
ADMBACl	Cationic		26	
B 1317	Anionic		40	
APES	Anionic		44	
Gafac	Anionic		75	
SDS	Anionic		39	
S-74	Anionic		49	
Akypo	Anionic		48	
S-150	Anionic		63	
C16TAB	Cationic	150	86	[40]
TritonX-100	Nonionic		97	
CTAB	Cationic	86	10	[41]
SDS	Anionic	86	3	
Tween-80	Nonionic	86	8	

Table 2.
Summary of contact angles changes after surfactants treatment.

different surfactants—anionic (SDS), cationic (CTAB), and nonionic (TritonX-100). The authors showed that cationic surfactant (CTAB) was more effective than anionic (SDS) and non-ionic (TritonX-100) ones in terms of changing the wetting state of the carbonate surface. The phenomenon of wettability changing has been explained by taking into account the irreversible desorption of acids adsorbed onto carbonate surfaces by CTAB surfactants via electrostatic interactions. Notably, for nonionic surfactants, the mechanism of wettability alteration was explained by ion exchange and polarization of π-electrons, whereas for anionic surfactants, the main mechanism of wettability alteration was found to be via hydrophobic interactions between surfactant tail and adsorbed hydrophobic oil components [43].

However, although the evident effectiveness of surfactant flooding mechanisms in wettability alteration, an optimal surfactant that can be both technically and economically feasible has not been found yet. Indeed, the main challenge is surfactant adsorption or retention during the injection process in the reservoir. The unproductive loss of surfactant decreases its effectiveness to lower brine/oil IFT and changing wettability.

Furthermore, although anionic surfactants have been found to be very promising for IFT reduction at reservoir conditions, the value of their adsorption onto hydrophobic or mixed-wet carbonate surfaces was estimated to be higher than in sandstone reservoirs. For instance, it was shown [44] that the adsorption value of typical anionic surfactant onto sandstone and limestone samples equaled 0.03 mg/g rock and 0.21 mg/g rock, respectively. In contrast, the adsorption value of cationic surfactant onto carbonate rocks was calculated to be only 0.12 mg/g rock [44]. The high adsorption value of anionic surfactants in carbonate reservoirs can be explained by the existence of electrostatic interactions between positively charged rock surfaces and negatively charged surfactant heads. Moreover, in high salinity brines of 5% $CaCl_2$, $MgCl_2$, or NaCl, the adsorption of anionic surfactant has been observed to be even higher due to the increased positive zeta-potential of the carbonate surfaces [45].

Therefore, although surfactants have been regarded as promising surface-active agents and laboratory experiments demonstrated their potential in wettability alteration, their industrial applications are limited due to high retention and adsorption onto reservoir rocks. Subsequently, different additives, such as alkalis and nanoparticles, have been studied as sacrificial agents to surfactant molecules in order to decrease their loss and improve efficiency for field applications.

4. Recent advances in chemical EOR: application of nanoparticles as oil/brine IFT and wettability modifiers

Traditionally, surfactant flooding as chemical EOR method has been developed and applied in sandstone oil reservoirs due to its economic and technical effectiveness [7]. Contrary to this, developing carbonate oil reservoirs with surfactant flooding is limited because of high loss of surfactant, resulting in increased operational costs. Nevertheless, many modeling and experimental studies have shown that surfactant flooding in oil-wet carbonate reservoirs could be a promising way of enhancing oil recovery [17]. Several effective ways have been proposed in the literature in order to overcome the high surfactant loss in carbonate reservoirs.

Conventionally, so-called "sacrificial" agents, such as sodium carbonate, sodium bicarbonate, or polyacrylate have been used to decrease the adsorption of surfactants [46]. The popularity of alkali addition to surfactants stems from their ability to increase pH (>7–8) that lead to an alteration of surface charge from positive toward more negative, which in turn results in a decrease in electrostatic attraction of anionic surfactant molecules to negatively charged surfaces.

However, it should be pointed out that some carbonate reservoirs consist of anhydrites ($CaSO_4$) that can react with alkali and cause the precipitation of $CaCO_3$ following the reaction [47]:

$$CaSO_4 + Na_2CO_3 \rightarrow Na_2SO_4 + CaCO_3 \downarrow \quad (3)$$

Interestingly, it was observed that nonionic or cationic surfactants show less adsorption value on carbonate rocks in comparison with anionic ones [39]. Although, the adsorption value of these surfactants is low, they have been reported to show less effectiveness in terms of oil/brine IFT reduction due to their chemical structure and properties [45]. Therefore, there is still a need for studies of developing the optimal chemical mixtures consistin of surfactants and "sacrificial" agents that can significantly reduce IFT, and exhibit low adsorption values.

Recently, the application of nanoparticle dispersions has been proposed to be promising alternative agent instead of alkali for decreasing adsorption of surfactant onto carbonate surfaces [48]. In the last years, the interest of using nanoparticles for enhancing effectiveness of surfactant EOR has been rapidly growing, with many studies being carried out [48–51]. Nanofluids or nano-assisted chemical EOR is defined as an injection of fluids that consist of 1–100 nm nanoparticles in colloidal suspension.

Several groups of nanoparticles exist—magnetic (Fe_3O_4, etc.), metal and non-metal oxides (ZrO_2, TiO_2, SiO_2, Al_2O_3, ZnO, etc.), and metallic (Cu, Pt, Au, Ag, etc.) [52–54]. For EOR purposes, the most commonly studied groups of nanoparticles are metal and non-metal oxides due to their unique physical and chemical properties [55]. For instance, these nanoparticles have shown good tolerance to mono and divalent ions (brine) and high thermal stability [56]. In this regard, many experimental studies have been carried out with these types of nanoparticles in order to evaluate their influence on surfactant EOR.

The main mechanisms of nanoparticles as EOR agents include wettability alteration, water/oil IFT reduction, increasing viscosity of injected fluids, disjoining pressure effect, and preventing asphaltene precipitation. Indeed, as it was shown in work [57], the inclusion of nanoparticles enhanced surfactant properties by increasing the stability of surfactant solutions and by helping in the reduction of oil/brine IFT. Moreover, studies also suggest the nanoparticles reduce the volume of surfactant needed for EOR, and thus improve the project economy.

In recent years, several studies have been reported about the influence of different nanoparticles (SiO_2, ZnO_2, and Al_2O_3,) on water/air surface and brine/oil IFTs in mixture with surfactants [50, 58, 59]. However, the influence of nanoparticles on the interfacial layer remains uncertain, with some contradicting trends existed in the literature. For instance, Ravera et al. [60] demonstrated that the surface and IFTs of cationic surfactant upon addition of SiO_2 nanoparticles increased. On the contrary, Al-Anssari et al. [61] and Lan et al. [62] reported that the inclusion of a small amount of SiO_2 nanoparticles to cationic and anionic surfactants resulted in IFT decrease. Furthermore, these results were supported by a study [63], where IFT reduction was observed in the presence of anionic surfactant and high (10 wt.%) concentration of nanoparticles. Moreover, according to the results of Zargartalebi et al. [64], IFT between anionic surfactant solutions with small concentrations (1000 ppm) of hydrophobic or hydrophilic SiO_2 nanoparticles and hydrocarbons decreased significantly when the surfactant concentration did not exceed the CMC. However, contrary to this, in the work [59] only a slight IFT reduction between oil and surfactant solutions with 0.5 wt.% ZnO_2 nanoparticles was observed. While further increase of surfactant concentration showed no effect on IFT in a range of all nanoparticle concentrations tested [59].

Interestingly the addition of SiO_2 nanoparticles to non-ionic surfactant Tween 20 showed a significant reduction of IFT from 44 to 10 mN/m [65]. The IFT decrease from 39 to 17.5 mN/m has been also observed while studying the SiO_2/Fe_2O_3 nanocomposites resulting in an overall 31% OOIP improvement [66]. As can be seen, despite having a significant number of publications in this area, researchers worldwide remain inconclusive over the interfacial behavior of nanoparticles augmented surfactant injection fluids, and further research is required in this area.

Since the development of hydrophobic carbonate oil reservoirs is emerging, different types and combinations of nanoparticles have been tested as additives to surfactant solutions in order to alter wettability towards more water-wet and thus enhance oil recovery [50, 67]. The authors in [67] reported that the addition of SiO_2 nanoparticles to anionic surfactant (SDS) aided in the reduction of water contact angle on carbonate surfaces. Importantly, the effect of nanoparticle addition was

Type of chemicals	Chemicals	Contact angle, °		% OOIP	Ref.
		Before treatment	After treatment		
Nanoparticles	SiO_2 + DIW	54.8	57.7	2.9	[68]
	SiO_2 + Brine	12	40	28	[69]
	SiO_2 + Ethanol	55	78	23	[70]
	SiO_2 + Xanthan	86	20	7.81	[71]
Nanocomposites	Fe_2O_3/SiO_2	138	52	31	[72]
	Fe_3O_4/chitosan	127	92	10.8	[73]
	TiO_2/SiO_2/xanthan	135	45	19.3	[74]
	NiO/SiO_2	174	32	—	[75]
Surfactant + Nanoparticles	L-Arg + SiO_2	141	57	13.1	[76]
	SDS + SiO_2	132	61	—	[67]
	L-Cys + SiO_2	141	48	12.7	[76]
	CTAB + NiO	150	60	—	[40]
	Triton + NiO	150	75	—	
	CTAB + ZrO_2	150	48	—	
	Triton + ZrO_2	150	78	—	
	Fatty acid methyl ester sodium sulfonate + SiO2	95	7.4	—	[77]

Table 3.
The summary of nanoparticles and/or nanocomposites effect on the contact angle of water on carbonate surfaces.

more pronounced when the surfactant's concentration was near CMC. Indeed, it was observed that the water advancing contact angle changed from ∿140° to 72° when only 0.2 wt.% of SiO_2 nanoparticles were added to the SDS surfactant solution. Whereas, the treatment in surfactant solution without nanoparticles led to less contact angle reduction ∿150° to 110°, illustrating oil-wet preference of carbonate surfaces.

Therefore, the authors [67] suggested that SiO_2 augmented surfactant solutions could be an effective fluid for EOR application in carbonate reservoirs, where the oil recovery process depends on wettability alteration. The effect of different nanoparticles and/or nanocomposites on surfactant property to change the wettability of carbonate surfaces has also been studied in many articles. The results of these studies are summarized in **Table 3**.

As it can be seen in **Table 3**, the inclusion of nanoparticles and/or nanocomposites to surfactant solutions helps in wettability alteration of carbonate rocks toward more hydrophilic. As a result, the oil recovery factor also increases. These results have been obtained in different studies with different types of nanoparticles tested, and thus, nanoparticles have been widely regarded as a promising EOR agent.

Therefore, the application of nano-assisted surfactant flooding may be a new chapter of chemical flooding for developing carbonate reservoirs. However, in order to scale this technology from the laboratory to field applications, more laboratory and modeling studies are required.

5. Field applications and challenges of chemical enhanced oil recovery

Generally, after many laboratory screening tests (stability, IFT, static and dynamic adsorption, wettability, and core flooding), successful candidates are selected for a single well tracer test (SWTT). In this test, the surfactant is injected into a well as a slug, and the oil saturation before and after is calculated. This test is performed in order to evaluate the amount of oil that can be reduced. Such a test is less expensive and usually is carried out before a field pilot test.

The scheme of injection may be different depending on reservoir conditions. As such, a scheme may include a so-called preflush, followed by a main slug and postflush. The preflush with alkaline is used to dilute the reservoir brine in order to reduce the concentration of divalent ions that can cause unfavorable surfactant degradation. For instance, sodium silicate, sodium carbonate, and sodium hydroxide were used in preflush slug in Bell Creek project and Salem field [78, 79]. As a postfluch slug usually includes polymers (biopolymers or synthetic) for improving the sweep efficiency after surfactant flooding.

Traditional types of surfactants used in EOR include but are not limited to petroleum sulfonate, ethyl sulfate, alkyl benzene sulfonate, carboxylates, etc. Moreover, many new surfactants are being synthesized mainly for EOR applications in high salinity and temperature conditions, such as biosurfactants and Gemini [80]. However, even if a novel surfactant shows promising economic and technical results in the laboratory, there is no guarantee that this surfactant will have the same effect in the field. The main reasons for this are surfactant production in a field-scales (tones), logistics, and the high cost of chemicals used for synthesis. For example, there are many works dedicated to the development of new effective surfactants that would be applicable in high temperatures and salinities conditions [17, 81]. However, when it comes to the field, the economic evaluation limits their implications.

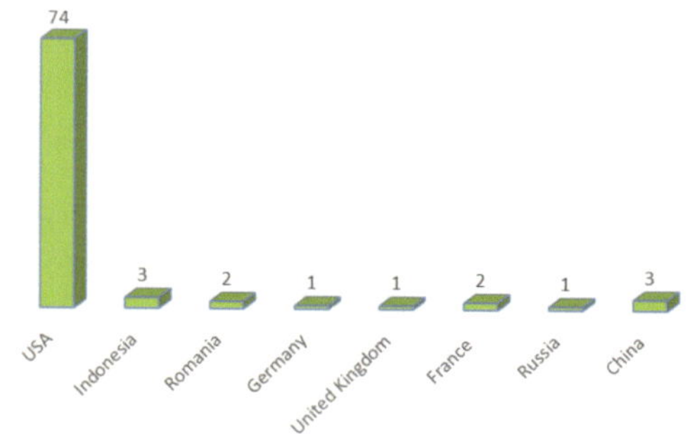

Figure 1.
Surfactant flooding projects that were conducted worldwide from the 1990s to 2000s [82].

Therefore, there are not many actual field projects reported in the literature. Moreover, historically, surfactant flooding as an EOR method was developed for sandstone oil reservoirs [17]. This stems from many factors, including pore matrix structure, mild reservoir conditions, low chemicals retention, and adsorption values. Carbonate reservoirs are considered to be promising candidates for surfactant EOR, but exhibit more complex structure and physical-chemical properties than sandstones, and thus only a few projects were conducted with them. **Figure 1** illustrates the number of surfactant projects conducted worldwide from the 1990s to the 2000s. It can be seen that the number for sandstone reservoirs surpasses the number of carbonates [82]. Only two projects were performed in carbonates in the USA—Cottonwood Creek [83, 84] and Yates field [85, 86], and one project in Semoga field Indonesia [87] However, it is interesting to point out that pilot tests of surfactant injection in carbonate reservoirs gave promising results. For instance, the Yates field pilot test showed a two-fold increase in oil recovery factor by using commercial surfactant Shell 91-8 [84, 86]. Therefore, surfactant flooding has been regarded as a promising alternative to CO_2 injection in carbonates [88].

It is important to note that in order to increase the applicability of surfactant EOR, the oil price should not be less than 50$/bbl and the cost of used chemicals should be decreased to a minimum so that the economy of the project will be profitable. This can be done, for instance, by developing the production factories close to the field, which will improve the local chemical production services. Moreover, the governmental support of local development companies is also needed to compensate for the economic risks of chemical EOR.

6. Conclusion

This chapter presents recent trends in chemical EOR with the emphasis on surfactant flooding and its applications for ensuring cost-effective hydrocarbons production. The mechanism of EOR applications and recent progress in chemical flooding have been addressed. The main challenges of chemical EOR have also been discussed. Field applications of surfactant EOR have been surveyed worldwide, illustrating a trend towards sandstone reservoirs rather than carbonates. Furthermore, a new type of chemical flooding, namely nano-assisted EOR, has been discussed with regards to improving surfactant flooding effectiveness. Nevertheless, the application of this new method is limited to laboratory tests and pilot scales. This can be attributed to some uncertainties associated with technology economics (instability of oil prices), a lack of understanding of the short-term and long-term environmental impact of nanoparticles applications. Therefore, a few recommendations for the future research of chemical EOR can be highlighted:

1. More studies are required for assessing the effectiveness of nanoparticles in chemical EOR, including an understanding of wettability alteration mechanisms and the impact on foam stability at reservoir conditions. Economic models of nanoparticles applications should be evaluated and compared with other chemicals involved.

2. Evaluation of nanoparticles application in carbonate reservoirs. So far, conventional chemical flooding is not economically feasible in carbonates. Therefore, new reagents should be investigated for developing an efficient EOR technique yet cost-effective.

3. Studies are required to investigate the environmental impact of nanoparticles, including reservoir rocks and their possible transfer to underground water. Furthermore, the separation technology of nanoparticles from produced water should also be studied, as already existing membranes may not be efficient with a new method.

4. There is a lack of modeling of nanoparticles application in EOR for designing field implications. So far, modeling with conventional software is limited, and thus more studies are required to fill this gap.

Acknowledgements

This work was supported by the Ministry of Science and Higher Education of the Russian Federation under agreement No. 075-10-2022-011 within the framework of the development program for a world-class Research Center.

Author details

Anastasia Ivanova* and Alexey Cheremisin
Skolkovo Institute of Science and Technology, Moscow, Russia

*Address all correspondence to: anastasia.ivanova@skoltech.ru

References

[1] Klemme H, Ulmishek GF. Effective petroleum source rocks of the world: stratigraphic distribution and controlling depositional factors. AAPG Bulletin. 1991;**75**:1809-1851

[2] Alamooti AM, Malekabadi FK. Fundamentals of Enhanced Oil and Gas Recovery from Conventional and Unconventional Reservoirs. Elsevier Inc.; Gulf Professional Publishing. 2018:1-40. DOI: 10.1016/B978-0-12-813027-8.00001-110.1016/C2016-0-04615-6

[3] Lake LW. Enhanced Oil Recovery. Prentice Hall Inc.; Englewood Cliffs 1989;**224**(4649):159-186

[4] Hirasaki G, Miller C, Puerto M. Recent advances in surfactant EOR. SPE Journal. 2011;**16**(4):3-5

[5] Lakatos I. Role of chemical IOR/EOR methods in the 21st Century, 18th World Petroleum Congress. Johannesburg, South Africa: World Petroleum Congress; 2005. WPC-18-0883: 2005

[6] Kamal MS, Hussein IA, Sultan AS. Review on surfactant flooding: Phase behavior, retention, IFT and field applications. Energy and Fuels. 2007;**31**(8):7701-7720. DOI: 10.1021/acs.energyfuels.7b00353

[7] Sheng JJ. Status of surfactant EOR technology. Petroleum. 2015;**1**:97-105. DOI: 10.1016/j.petlm.2015.07.003

[8] Melrose JC. Role of capillary forces in determining microscopic displacement efficiency for oil recovery by waterflooding. Journal of Canadian Petroleum Technology. 2010;**13**:54-62. DOI: 10.2118/74-04-05

[9] Fulcher RA, Ertekin T, Stahl CD. Effect of capillary number and its constituents on two-phase relative permeability curves. Journal of Petroleum Technology. 1985;**37**:249-260. DOI: 10.2118/12170-PA

[10] Morrow NR. Interplay of capillary, viscous and buoyancy forces in the mobilization of residual oil. Journal of Petroleum Technology. 1979;**18**:35-46

[11] Howe AM, Clarke A, Mitchell J, Staniland J, Hawkes L, WhalanC.Visualisingsurfactantenhanced oil recovery. Colloids and Surfaces, A: Physicochemical and Engineering Aspects. 2015;**480**:449-461. DOI: 10.1016/j.colsurfa.2014.08.032

[12] Hou J, Liu Z, Zhang S, Yue XA, Yang J. The role of viscoelasticity of alkali/surfactant/polymer solutions in enhanced oil recovery. Journal of Petroleum Science and Engineering. 2005;**47**(3-4):219-235. DOI: 10.1016/j.petrol.2005.04.001

[13] Holmberg K, Jonsson B, Kronberg B, Lindman B. Surfactants and Polymers in Aqueous Solution. 2nd ed. New York: John Wiley & Sons; 2003. p. 562c

[14] Fanchi JR, Christiansen RL. Applicability of Fractals to the Description of Viscous Fingering. SPE Annual Technical Conference and Exhibition. San Antonio, Texas; 1989. DOI: 10.2118/19782-MS

[15] Daccord G, Lenormand R. Fractal patterns from chemical dissolution. Nature. 1987;**325**:41-43. DOI: 10.1038/325041a0

[16] Olajire AA. Review of ASP EOR (alkaline surfactant polymer enhanced

oil recovery) technology in the petroleum industry: Prospects and challenges. Energy. 2014;**77**:963-982. DOI:10.1016/j.energy.2014.09.005

[17] Sheng J. Modern Chemical Enhanced Oil Recovery: Theory and Practice. Amsterdam: Elsevier Publication Inc; 2011

[18] Wang H, Cao X, Zhang J, Zhang A. Development and application of dilute surfactant–polymer flooding system for Shengli oilfield //. Journal of Petroleum Science and Engineering. 2009;**65**:45-50

[19] Negin C, Ali S, Xie Q. Most common surfactants employed in chemical enhanced oil recovery. Petroleum. 2017;**3**:197-211. DOI: 10.1016/j.petlm.2016.11.007

[20] Abdallah W, Buckley JS, Carnegie J, Edwards J, Fordham BHE, Graue A, et al. Fundamentals of wettability. Oilfield Review. 2007;**19**:44-61

[21] Jadhunandan PP, Morrow NR. Effect of wettability on waterflood recovery for crude oil/brine/rock systems. In: Paper SPE 22597 prepared for presentation at the 66th Annual Technical Conference and Exhibition, Dallas, TX. 1991

[22] Ivanova A, Mitiurev N, Cheremisin A, Orekhov A, Kamyshinsky R, Vasiliev A. Characterization of organic layer in oil carbonate reservoir rocks and its effect on microscale wetting properties. Scientific Reports Nature. 2019;**9**:10667. DOI: 10.1038/s41598-019-47139-y

[23] Buckey JS, Liu Y, Monsterleet S. Mechanisms of wetting alteration by crude oils. SPE Journal. 1998;**3**:54-61. DOI: 10.2118/37230-pa

[24] Madsen L, Lind I. Adsorption of carboxylic acids on reservoir minerals from organic and aqueous phase. SPE Reservoir Evaluation and Engineering. 1998;**1**:47-51. DOI: 10.2118/37292-PA

[25] Thomas MM, Clouse JA, Longo JM. Adsorption of organic compounds on carbonate minerals: 1. Model compounds and their influence on mineral wettability. Chemical Geology. 1993;**109**:201-213. DOI: 10.1016/0009-2541(93)90070-Y

[26] Muriel H, Madland MV, Korsnes RI. Evolution of wetting index over time in mixed wet kansas chalk using triaxial cells [Thesis for Master of Science]. 2017. Available from: https://uis.brage.unit.no/uis-xmlui/handle/11250/2463444

[27] Amott E. Observations relating to the wettability of porous rock. Transactions of AIME. 1959;**216**:156-162

[28] Donaldson EC, Thomas RD, Lorenz PB. Wettability determination and its effect on recovery efficiency. Journal of Society of Petroleum Engineers. 1969;**9**:13-20. DOI: 10.2118/2338-PA

[29] Donaldson EC. Oil-water-rock wettability measurement. ACS Symposium Chemical Engineering Combustion Deposit Conference; Atlanta, USA; 1981

[30] Yuan Y, Lee TR. Contact angle and wetting properties. Journal of Surface Science and Technology. 2013;**51**:3-34. DOI: 10.1007/978-3-642-34243-1_1

[31] Tul'bovich BI. Method for Determination of Wettability in Hydrocarbon-Bearing Rocks: OST 39-180-85. Moscow: Nedra; 1979

[32] Anderson W. Wettability literature survey-part 2: wettability measurement. Journal of Petroleum Technology. 1986;**38**:1246-1262. DOI: 10.2118/13933-PA

[33] Kowalewski E, Boassen T, Torsaeter O. Wettability alterations due to aging in crude oil; wettability and Cryo-ESEM analysis. Journal of Petroleum Science and Engineering. 2003;**39**:377-388. DOI: 10.1016/S0920-4105(03)00076-7

[34] Robin M, Combes R, Degreve F, Cuiec L. Wettability of porous media from environmental scanning electron microscopy: From model to reservoir rocks. SPE International Conference on Oilfield Chemistry. 1997;37235:18-21 DOI: 10.2118/37235-MS

[35] Park J, Han HS, Kim YC, Ahn JP, Ok MR, Lee KE, et al. Direct and accurate measurement of size dependent wetting behaviors for sessile water droplets. Scientific Reports. 2015;**5**:1-13. DOI: 10.1038/srep18150

[36] Ensikat H, Schulte AJ, Koch K, Barthlott W. Droplets on super hydrophobic surfaces: visualization of the contact area by cryo-scanning electron microscopy. Langmuir. 2009;**25**:13077-13083. DOI: 10.1021/la9017536

[37] Deglint H, Clarkson CR, DeBuhr C, Ghanizadeh A. Live imaging of micro-wettability experiments performed for low-permeability oil reservoirs. Scientific Reports. 2017;**7**:1-13. DOI: 10.1038/s41598-017-04239-x

[38] Ivanova A, Orekhov A, Markovic S, Iglauer S, Grishin P, Cheremisin A. Live imaging of micro and macro wettability variations of carbonate oil reservoirs for enhanced oil recovery and CO_2 trapping/storage. Scientific Reports. 2022;**12**:1262. DOI: 10.1038/s41598-021-04661-2

[39] Standnes DC, Austad T. Wettability alteration in chalk. Society of Petroleum Engineers. 2000;**28**:123-143. DOI: 10.1016/s0920-4105(00)00084-x

[40] Nwidee LN, Lebedev M, Barifcani A, Sarmadivaleh M, Iglauer S. Wettability alteration of oil-wet limestone using surfactant-nanoparticle formulation. Journal of Colloid and Interface Science. 2017;**504**:334-345. DOI: 10.1016/j.jcis.2017.04.078

[41] Kumar S, Mandal A. Studies on interfacial behavior and wettability change phenomena by ionic and nonionic surfactants in presence of alkalis and salt for enhanced oil recovery. Applied Surface Science. 2016;**372**:42-51. DOI: 10.1016/j.apsusc.2016.03.024

[42] Zhang DL, Liu S, Puerto M, Miller CA, Hirasaki GJ. Wettability alteration and spontaneous imbibition in oil-wet carbonate formations. Journal of Petroleum Science and Engineering. 2006;**52**:213-226. DOI: 10.1016/j.petrol.2006.03.009

[43] Jarrahian K, Seiedi O, Sheykhan M, Vafaie Sefti M, Ayatollahi S. Wettability alteration of carbonate rocks by surfactants: A mechanistic study. Colloids and Surfaces A: Physicochemical and Engineering Aspects. 2012;**410**:1-10. DOI: 10.1016/j.colsurfa.2012.06.007

[44] Mannhardt K, Schramm LL, Novosad JJ. Effect of Rock Type and Brine Composition on Adsorption of Two Foam-Surfactants. In: Paper SPE 20463 presented at the 1990 SPE Annual Technical Conference and Exhibition. 1990

[45] Jian G, Puerto M, Wehowsky A, Miller C, Hirasaki GJ, Biswal SL. Characterizing adsorption of associating surfactants on carbonates surfaces. Journal of Colloid and Interface Science. 2018;**513**:684-692. DOI: 10.1016/j.jcis.2017.11.041

[46] ShamsiJazeyi H, Verduzco R, Hirasaki GJ. Reducing adsorption of

anionic surfactant for enhanced oil recovery: Part II. Applied aspects. Colloids and Surfaces A: Physicochemical and Engineering Aspects. 2014;**453**:168-175. DOI: 10.1016/j.colsurfa.2014.02.021

[47] Mohnot SM, Bae JH, Foley WL. A study of mineral/alkali reactions. Society of Petroleum Engineers. 1987;**2**:653-663

[48] Ivanova AA, Cheremisin AN, Spasennykh MY. Application of nanoparticles in chemical EOR. EAGE Conference Proceedings. 2017;**2017**:1-10. DOI: 10.3997/2214-4609.201700247

[49] Cheraghian G, Hendraningrat L. A review on applications of nanotechnology in the enhanced oil recovery part a: Effects of nanoparticles on interfacial tension. International Nano Letters. 2016;**6**:129-138. DOI: 10.1007/s40089-015-0173-4

[50] Ogolo NA, Olafuyi OA, Onyekonwu MO. Enhanced oil recovery using nanoparticles. Society of Petroleum Engineers. 2012;**160847-MS**:1-9. DOI: 10.2118/160847-MS

[51] Nettesheim F, Liberatore MW, Hodgdon TK, Wagner NJ, Kaler EW, Vethamuthu M. Influence of nanoparticle addition on the properties of wormlike micellar solutions. Langmuir. 2008;**24**:7718-7726. DOI: 10.1021/la800271m

[52] Morrow L, Potter DK, Barron AR. Detection of magnetic nanoparticles against proppant and shale reservoir rocks. Journal of Experimental Nanoscience. 2015;**10**:1028-1041. DOI: 10.1080/17458080.2014.951412

[53] White RJ, Luque R, Budarin VL, Clark JH, Macquarrie DJ. Supported metal nanoparticles on porous materials, methods and applications. Chemical Society Reviews. 2009;**38**:481-494. DOI: 10.1039/B802654H

[54] Fedele L, Colla L, Bobbo S, Barison S, Agresti F. Experimental stability analysis of different water-based nanofluids. Nanoscale Research Letters. 2011;**6**:1-8. DOI: 10.1186/1556-276X-6-300

[55] Hendraningrat L, Torsæter O. Metal oxide-based nanoparticles: Revealing their potential to enhance oil recovery in different wettability systems. Applied Nanoscience. 2015;**5**:181-199. DOI: 10.1007/s13204-014-0305-6

[56] Yu W, France DM, Routbort JL, Choi SU. Review and comparison of nanofluid thermal conductivity and heat transfer enhancements. Heat Transfer Engineering. 2009;**29**:432-460. DOI: 10.1080/01457630701850851

[57] Sun X, Zhang Y, Chen G, Gai Z. Application of nanoparticles in enhanced oil recovery: A critical review of recent progress. Energies. 2017;**10**:1-33. DOI: 10.3390/en10030345

[58] Esmaeilzadeh P, Fakhroueian Z, Bahramian A, Arya S. Influence of ZrO2 nanoparticles including SDS and CTAB surfactants assembly on the interfacial properties of liquid-liquid, liquid-air and liquid-solid surface layers. Journal of Nanoparticle Research. 2012;**21**:15-21. DOI: 10.4028/www.scientific.net/jnanor.21.15

[59] Esmaeilzadeh P, Hosseinpour N, Bahramian A, Fakhroueian Z, Arya S. Effect of ZrO2 nanoparticles on the interfacial behavior of surfactant solutions at air-water and n-heptane-water interfaces. Fluid Phase Equilibria. 2014;**361**:289-295. DOI: 10.1016/j.fluid.2013.11.014

[60] Ravera F, Santini E, Loglio G, Ferrari M, Liggieri L. Effect of

nanoparticles on the interfacial properties of liquid/liquid and liquid/air surface layers. The Journal of Physical Chemistry. B. 2006;**110**:19543-19551. DOI: 10.1021/jp0636468

[61] Al-Anssari S, Wang S, Barifcani A, Iglauer S. Oil-water interfacial tensions of silica nanoparticle-surfactant formulations. Tenside, Surfactants, Detergents. 2017;**54**:334-341. DOI: 10.3139/113.110511

[62] Lan Q, Yang F, Zhang S, Liu S, Xu J, Sun D. Synergistic effect of silica nanoparticle and cetyltrimethyl ammonium bromide on the stabilization of O/W emulsions. Colloids and Surfaces A: Physicochemical and Engineering Aspects. 2007;**302**:126-135. DOI: 10.1016/J.COLSURFA.2007.02.010

[63] Ma H, Luo M, Dai LL. Influences of surfactant and nanoparticle assembly on effective interfacial tensions. Physical Chemistry Chemical Physics. 2008;**10**:2207-2213. DOI: 10.1039/b718427c

[64] Zargartalebi M, Barati N, Kharrat R. Influences of hydrophilic and hydrophobic silica nanoparticles on anionic surfactant properties: Interfacial and adsorption behaviors. Journal of Petroleum Science and Engineering. 2015;**119**:36-43. DOI: 10.1016/j.petrol.2014.04.010

[65] Biswal NR, Singh JK. Interfacial behavior of nonionic Tween 20 surfactant at oil–water interfaces in the presence of different types of nanoparticles. RSC Advances. 2016;**6**:113307-113314. DOI: 10.1039/c6ra23093h

[66] Kazemzadeh Y, Dehdari B, Etemadan Z, Riazi M, Sharifi M. Experimental investigation into Fe3O4/SiO2 nanoparticle performance and comparison with other nanofluids in enhanced oil recovery. Petroleum Science. 2019;**16**:578-590. DOI: 10.1007/s12182-019-0314-x

[67] Al-Anssari S, Nwidee LN, Arif M, Wang S, Barifcani A, Lebedev M, et al. Wettability alteration of carbonate rocks via nanoparticle-anionic surfactant flooding at reservoirs conditions. SPE Symposium: Production Enhancement and Cost Optimisation. 2017;**189203-MS**:1-12. DOI: 10.2118/189203-ms

[68] Bayat EA, Junin R, Samsuri A, Piroozian A, Hokmabadi M. Impact of metal oxide nanoparticles on enhanced oil recovery from limestone media at several temperatures. Energy & Fuels. 2014;**28**:6255-6266. DOI: 10.1021/ef5013616

[69] Li Y, Dai C, Zhou H, Wang XLVW, Zhao M. Investigation of Spontaneous Imbibition by Using a Surfactant-Free Active Silica Water-Based Nanofluid for Enhanced Oil Recovery. Energy & Fuels. 2017;**32**:287-293. DOI: 10.1021/acs.energyfuels.7b03132

[70] Roustaei A, Moghadasi J, Bagherzadeh H, Shahrabadi A. An experimental investigation of polysilicon nanoparticles' recovery efficiencies through changes in interfacial tension and wettability alteration. Society of Petroleum Engineers International Oilfield Nanotechnology Conference Exhibition. 2012;**156976**:1-7 DOI: 10.2118/156976-MS

[71] Saha R, Uppaluri RV, Tiwari P. Silica nanoparticle assisted polymer flooding of heavy crude oil: emulsification, rheology, and wettability alteration characteristics. Industrial and Engineering Chemistry Research. 2018;**57**:6364-6376. DOI: 10.1021/acs.iecr.8b00540

[72] Kazemzadeh Y, Sharifi M, Riazi M, Rezvani H, Tabaei M. Potential effects of metal oxide/SiO2 nanocomposites in EOR processes at different pressures. Colloids and Surfaces A: Physicochemical and Engineering Aspects. 2018;**559**:372-384. DOI: 10.1016/j.colsurfa.2018.09.068

[73] Rezvani H, Riazi M, Tabaei M, Kazemzadeh Y, Sharifi M. Experimental investigation of interfacial properties in the EOR mechanisms by the novel synthesized Fe3O4/Chitosan nanocomposites. Colloids and Surfaces A: Physicochemical and Engineering Aspects. 2018;**544**:15-27. DOI: 10.1016/j.colsurfa.2018.02.012

[74] Ali JA, Kolo K, Manshad AK, Stephen KD. Potential application of low-salinity polymeric-nanofluid in carbonate oil reservoirs: IFT reduction, wettability alteration, rheology and emulsification characteristics. Journal of Molecular Liquids. 2019b;**284**:735-747. DOI: 10.1016/j.molliq.2019.04.053

[75] Dahkaee KP, Sadeghi MT, Fakhroueian Z, Esmaeilzadeh P. Effect of NiO/SiO2 nanofluids on the ultra-interfacial tension reduction between heavy oil and aqueous solution and their use for wettability alteration of carbonate rocks. Journal of Petroleum Science and Engineering. 2019;**176**:11-26. DOI: 10.1016/j.petrol.2019.01.024

[76] Asl HF, Zargar G, Manshad AK, Takassi MA, Ali JA, Keshavarz A. Effect of SiO_2 nanoparticles on the performance of L-Arg and L-Cys surfactants for enhanced oil recovery in carbonate porous media. Journal of Molecular Liquids. 2020;**300**:1-12. DOI: 10.1016/j.molliq.2019.112290

[77] Scerbacova A, Ivanova A, Mukhina E, Ushakova A, Bondar M, Cheremisin A. Screening of surfactants for huff-n-puff injection into unconventional reservoirs. Paper presented at the SPE Russian Petroleum Technology Conference, Virtual. 2021;**206431-MS**:1-14. DOI: 10.2118/206431-MS

[78] Johnson RE, Kennelly RG, Schwarz JR. Chemical and physical applications in treating produced-fluid emulsions of the Bell creek micellar/polymer flood. Society of Petroleum Engineers. 1988;**3**(2):210-216

[79] Whiteley RC, Ware JW. Low-tension waterflood pilot at the Salem unit, Marion County, Illinois Part 1: field implementation and results. Journal of Petroleum Technology. 1977;**29**(8):925-932

[80] Wang H-Z, Liao G-Z, Song J. Combined chemical flooding technologies. In: Shen PP, editor. Technological Developments in Enhanced Oil Recovery. Petroleum Industry Press; 2006. pp. 126-188

[81] Yuan CD, Pu WF, Wang XC, Sun L, Zhang YC, Cheng S. Effects of interfacial tension, emulsification, and surfactant concentration on oil recovery in surfactant flooding process for high temperature and high salinity reservoirs. Energy & Fuels. 2015;**29**:6165-6176. DOI: 10.1021/acs.energyfuels.5b01393

[82] Sheng JJ. Review of surfactant enhanced oil recovery in carbonate reservoirs. Advances in Petroleum Exploration and Development. 2013;**6**:1-10

[83] Weiss WW, Xie X, Weiss J, Subramanium V, Taylor A, Edens F. Artificial intelligence used to evaluate 23 single-well surfactant-soak treatments. Society of Petroleum Engineers. 2006;**9**:209-216. DOI: 10.2118/89457-PA

[84] Xie X, Weiss WW, Tong Z, Morrow NR. Improved oil recovery from carbonate. SPE Journal. 2005;**10**:276-285. DOI: 10.2118/89424-PA

[85] Chen HL, Lucas LR, Nogaret LAD, Yang HD, Kenyon DE. Laboratory monitoring of surfactant imbibition with computerized tomography. Society of Petroleum Engineers. 2000;**4**:16-25. DOI: 10.2118/59006-MS

[86] Yang HD, Wadleigh EE. Dilute surfactant IOR–design improvement for massive, fractured carbonate applications. In: Paper SPE 59009 presented at the SPE International Petroleum Conference and Exhibition. 2000. DOI: 10.2118/59009-MS

[87] Rilian NA, Sumestry M, Wahyuningsih W. surfactant stimulation to increase reserves in carbonate reservoir "a case study in Semoga field". In: Paper SPE 130060 presented at the SPE EUROPEC/EAGE Annual Conference and Exhibition. 2010. DOI: 10.2118/130060-MS

[88] Manrique EJ, Muci VE, Gurfinkel ME. EOR field experiences in carbonate reservoirs in the united states. Society of Petroleum Engineers. 2007;**10**: 667-686. DOI: 10.2118/100063-PA

Section 4

Polymers Flooding

Chapter 8

Rock Lithology-Based Laboratory Protocols and Best Practices for Polymer Screening for EOR Application in Conventional and Harsh Reservoirs

Yani Araujo and Mariela Araujo

Abstract

EOR applications have seen a recent shift in focus from onshore use in sandstone formations to more complex environments, such as carbonates and reservoirs in offshore settings. This explains the attention given thus far to develop EOR screening criteria mainly for use in sandstone reservoirs, where significant success has been observed. Screening of chemical EOR applications is mostly centered on the evaluation and characterization of fluids and some formation properties without explicit consideration of the formation lithology in some cases. Standardized lithology-based laboratory protocols are required to avoid cost overruns on screening and the design of fluid formulations for specific use in a particular reservoir. Such is the case of carbonates, typically highly heterogenous formations often found in high salinity and temperature conditions, where most available chemicals have limited applicability, and where standard screening protocols are not reliable or lack proper detection limits. In this chapter, we present an integrated laboratory workflow for polymer screening with recommended protocols based on formation lithology. It was derived from successful polymer application in the O&G industry and our own experimental work over the last two decades. Its use results in better quality data with time savings, contributing to a better understanding of the field application.

Keywords: polymer screening, lithology, lab protocol, best practices, reservoir conditions

1. Introduction

Most EOR screening criteria focus on sandstone onshore applications, where the cost of implementation allows for their use. As a result, extensive experience has been gained through the years in those reservoirs. More recently, the attention has moved to more complex environments, such as onshore carbonates and reservoirs in offshore settings, where the size of the price seems attractive enough to consider EOR as part of the field development [1, 2].

For a successful implementation of chemical EOR in different types of lithologies and depositional settings, standardized laboratory protocols are required to avoid cost overruns on screening, and the design of fluid formulations for specific use in a particular reservoir. The screening of polymers for use in EOR application for carbonates is particularly challenging since they are highly heterogenous formations frequently found in relatively high temperature and high salinity conditions, where most of the available chemicals for EOR have limited applicability. The standard protocols for screening are not reliable due to the typical detection limits making it difficult to understand which formulations work best in those complex environments.

Even though there is extensive literature on polymer applications for EOR, it is noticed that there is no consistency in the way polymers are screened in the laboratory for their use, making it difficult to even compare products from different vendors [3]. Some of the tests do not consider the formation lithology and the potential influence of mineralogy on the polymer screening process. These elements reflect a clear need for an integrated workflow for polymer laboratory screening that explicitly considers the rock lithology and reservoir conditions. Such a workflow would be very relevant for applications in reservoirs exposed to high temperature and high salinity.

In this chapter, we provide an overview of the key steps of a laboratory workflow and the recommended protocols for the laboratory measurements according to the formation lithology where the polymer will be used. The workflow has been derived based on the successful application of polymers in the oil and gas industry, and our own experimental work over the last two decades. The use of an integrated workflow with standardized laboratory protocols helps with the screening process not only avoiding repeating tests that frequently are inconclusive, but also provides better quality data saving time and allowing for a better understanding of the potential field application.

The different steps of the proposed workflow are presented in detail and illustrated with the use of reservoir samples to demonstrate the limitations of the standard experimental protocols used, and a discussion on the relevant best practices industry wide from successful applications in the field. The emphasis is given to the need to move away from "model" lab conditions by using reservoir rock and fluid samples as best practices, the recommendation to execute the tests at reservoir pressure and temperature conditions on restored cores, the use of representative saturation conditions, and the value of considering a variety of potential testing scenarios that reflect the expected changes that the polymer solutions will experience during their use in the field application.

Polymers have been used to increase the areal and vertical sweep efficiency of *in situ* hydrocarbons in the process of polymer flooding to improve recovery. Due to the high cost of chemicals and trending low oil prices, it is essential to optimize polymer flooding strategies to shorten the polymer injection time and also maximize efficiency to keep the operation economically feasible, especially in carbonate and offshore oil fields.

Despite having many successful polymer floods [3], field application of polymer flooding still faces several challenges, in particular for offshore EOR projects:

1. Reservoir and Wells: Chemicals suitable for different reservoir lithologies, presence of heterogeneities such as fractures/faults, high-permeability layers (as in Bohai [1]), high-salinity formation water, high-viscosity crude oil, distribution of residual oil distribution, large well spacing, and optimal well patterns [1–3].

2. Water source: Use of freshwater versus high salinity water [1].

3. Available working space: typically, there is limited space in the offshore platform for the equipment to prepare polymer solution, and the injection system needs to be compact, flexible, and efficient [1].

4. Environmental requirements that for offshore implementations are high as the produced fluid from polymer flooding needs to meet the standards set by government/regulatory agencies [1, 3].

There is a huge difference between polymer injection in offshore and onshore fields including the timeline, type of equipment to measure critical parameters at high temperature and pressure, and the lack of laboratory standards to evaluate the performance of polymers for offshore application [1]. In the last two decades we have seen the result from R&D work, field trials, and applications of offshore chemical EOR techniques translated into equipment and techniques that enabled offshore EOR, including high-efficiency driving agents, platform polymer injection distribution systems, produced fluid treatment technology, and a variety of methods for performance evaluation of early-stage polymer flooding [1, 4].

2. Types of polymers for EOR applications

Even though there are different types of polymers available in the market for EOR applications, including powder, liquid, and emulsion polymers, selecting the right type according to the reservoir lithology, properties, and conditions is not an easy task.

The most common polymers used in field applications are anionic polyacrylamide, hydrolyzed polyacrylamide (HPAM), and co-polymers, such as sodium acrylate. The addition of functional monomers is a way to modify them for specific applications. For example, N-vinyl pyrrolidone has been proven effective to improve the thermal stability and salt tolerance of conventional polymers [4, 5]. Biopolymers are also used; however, they have limitations due to environmental concerns and operational costs related to degradation by bacteria resulting in the need for frequent biocide treatment. Schizophyllan is in use in the Bockstedt field in the North Sea since 2006 for its stability for temperatures up to 130°C while being environmentally friendly [6]. Other polymers, such as the one proposed by Skauge [7] for use at low concentration and not field tested yet, are generated by a crosslinking reaction between aluminum and HPAM.

Chang [8] reported results from an evaluation of polymer flood in 16 USA field cases with only one done on a limestone formation where polyacrylamide polymers and Xanthan Gum were used. Literature data on 72 polymer field applications, including six offshore projects, show that most projects used HPAM (92%) and biopolymers. Only one project used an associative polymer in Bohai Bay, China [9]. The emulsion type of HPAM has been the preferred way of transportation except for some fields (Dalia, Bohai and SZ36-1) where the powder form was used.

Recent review work summarizes the available polymers for use in high temperature and high salinity reservoirs [10, 11]. These polymers are mainly based on acrylamide. Modified acrylamide co-polymers and polyacrylamides have shown better tolerance to high salinity and temperature than HPAMs. For example, sulfonate-based acrylamide polymer works well for temperatures up to 105°C.

Associative polymers seem to be an option for high salinity environments [12, 13]. A promising candidate for use in carbonates is the synthetic polymer N-vinylpyrrolidone-free with acrylamide tertiary butyl sulfonic acid, which shows thermal stability up to 140°C and salinity tolerance up to 220 g/l [14].

Candidate polymers for EOR applications should meet the following requirements [3, 11]—(a) high viscosity at reservoir conditions, (b) low impact on injectivity—thus good solubility and filterability, (c) stability at reservoir conditions (temperature, salinity, pressure, and hardness), (d) compatibility with injection chemicals and *in situ* fluids, (e) low adsorption, (f) friendly to the environment (including CO_2 footprint), (g) low impact on water separation processes.

The actual presentation of the polymer, either in powder or emulsion form is not so critical for the actual field application, since the polymer chemistry can be achieved in any of these presentations. What is critical is the screening process they follow to determine the most suitable candidates.

The following points have been reported as relevant when selecting the polymer presentation:

a. Polymers in powder form are popular for onshore use, whereas liquid polymers are considered for use in remote locations, such as offshore reservoirs and more hostile environments

b. Preparation of the emulsion polymer requires more attention than its use in powder form since it behaves like a multi-component system [15].

c. The polymer solution from powder is prepared on-site, thus it is easier to use compared to the emulsion type of polymer [15, 16].

d. The equipment footprint for dissolution of emulsion polymers is smaller than the required for powders. Its transportation is also easier. In general, it is challenging to use powders for offshore applications [15].

e. Since the polymer powders are designed to cover a wider range of conditions, there is no need to adapt the formulation to the actual reservoir conditions. In the case of the emulsion type, the surfactant package must be adapted to the reservoir brine characteristics (temperature and salinity) to allow a perfect release of the macromolecules and a good dispersion of the oil droplets avoiding injectivity issues [17].

f. Emulsion polymers require less time to dilute compared to powders (10 min by phase inversion for emulsions vs. 120 min for powders) [17].

g. Powder polymers seem to be more thermal stable than emulsion polymers [17].

h. Emulsion type of polymers has a lower active concentration range (~30–48%) requiring larger volumes for transportation. Powder forms active concentration can be up to 100%, thus lower volumes are required to be transported for field application [18].

i. In terms of oil recovery percentage and pressure response, no significant differences have been observed between both types of polymers [16, 18].

3. Workflow for laboratory screening

Following oil and gas best practices for laboratory polymer evaluation [3, 11, 15, 19–24], a consistent workflow is proposed to select polymers for EOR applications, as illustrated schematically in **Figure 1**.

The first step relates to the pre-selection of commercially available polymers. This is done based on a comprehensive list of key reservoir parameters to be shared with the polymer's providers. The minimum set of parameters required for the pre-selection of a polymer for lab screening includes reservoir temperature, rock lithology (carbonate, sandstone, shale, etc.), formation brine composition and properties (salinity, pH, total dissolved solids (TDS), R+), oil properties (API, Density, TAN and water content).

Once the pre-selected polymers are received in the lab, the polymers are prepared (dissolved in reservoir brine) (Step 2) and quality checked by evaluation of the amount of insoluble particles (<0.01%) and viscosity measurement at reservoir temperature (Step 3). Polymers that pass the Quality Check (QC) step are considered for the next screening steps (Steps 4 and 5). It is recommended that polymer characterization and performance evaluation to be done by an independent laboratory and not by the polymer provider.

In the next sub-sections, the recommended tests are presented and discussed.

3.1 Preparation of polymer solutions

Polymer transport and preparation can be challenging for a chemical EOR project. Lab analysis can help to support the field implementation. The preparation of all-polymer stock solutions, regardless of polymer type, follows the well-known best practices in the industry [15, 19–22].

Powder hydrolyzed polyacrylamide (HPAM) and polyacrylamides (PAM) type of polymers stock solutions are prepared at a concentration of 5000 ppm. It is recommended the solutions be prepared inside a glove box previously evacuated and purged with nitrogen (three times) to remove all oxygen. With the brine under

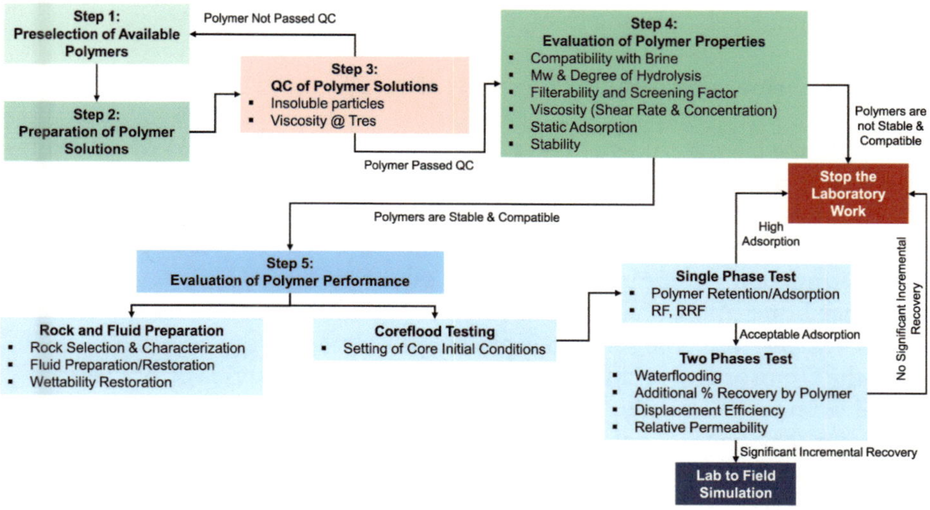

Figure 1.
Overview of the proposed workflow for polymer screening.

stirring at 400 rpm with an overhead mixer, the polymer is slowly introduced to avoid the formation of fisheyes (over a period of 30 s). The mixture is stirred for 5 min at full vortex. Additional stirring up to 120 min is required in some cases for full solution homogenization.

In the case of the emulsion type polymers according to the mechanical dissolution conditions, the time to dissolve it may vary from a few seconds to a few minutes. Emulsions are usually most effective when the dissolving operation is carried out continuously. It is important to observe a final concentration in a solution of approximately 5 g/l of active material. The emulsion solution is injected using a syringe into the brine, and the entire emulsion volume is added as quickly as possible. After 10 min, the solution should be homogeneous with light color and viscous appearance.

As a protective measure, additional chemicals can be added to prevent polymer degradation due to the presence of oxygen and iron, and bacteria. If that is the case, it is important to verify the final solution is properly homogenized [15, 22].

Polymer mixing has been recognized by operators as an important step in polymer screening, and specific laboratory mixing protocols have been proposed for use with liquid polymers. The wellness of the mixing is described by the specific mixing energy parameter [25] that helps to upscale data from lab to field. It can be also applied for the mixing of powder polymers.

3.2 Polymer stock solution filtration and dilution

The polymer stock solution must be filtered before its use to remove any aggregates present. It is recommended to filter the stock solution through a 5 μm polycarbonate filter. In most cases, polymers are filtered using a high-pressure filter assembly. A pressure differential of 15 psi is typically obtained when using nitrogen or argon. Once the filtration step is complete, the remaining fluid is discarded.

Modified filtration methods are required when working with low permeability carbonate rocks before making rheology and injectivity measurements [26].

To achieve the desired polymer concentration, the next step is to dilute the prepared and filtered stock solution. From the stock solution in the globe box, brine is gradually introduced, and the solution is continuously stirred for another hour or so to completely homogenize the solution.

3.3 Polymer-brine compatibility

Step 4 of the laboratory screening workflow for both carbonates and sandstones formations regardless of reservoir conditions starts with the evaluation of the compatibility between the polymer and reservoir brines (formation and injection). The best practice is to do the test in both brine. If there is a need to speed up the study, the test with formation water can be skipped for sandstones [11, 22].

Polymer-brine compatibility is evaluated at a fixed solution concentration and reservoir temperature. The solution is placed into a temperature-resistant bottle in an oven at the required temperature for a period of at least 21 days. The bottles are removed from the oven daily. The appearance of the solution is observed, and the viscosity for the corresponding bottle is measured. Any change should be registered.

For quality assurance purposes it is recommended to do the test on duplicate samples. If the results are not similar for both samples, the procedure must be repeated. After mixing, the solutions are visually analyzed for solubility and classified using the following criteria:

- Compatible, Clear, or Soluble (C): Transparent clear single-phase solution observed. Changes in Viscosity <1% over a period of 21 days.

- Non-Compatible or Insoluble (NC): Two distinct phases or precipitates are observed.

Figure 2 provides an example of compatibility for three different polymers dissolved in the same brine and aged at the same temperature. The solution with the best solubility is the one that shows the smaller change in viscosity over a period of 21 days.

3.4 Polymer solution properties

After the brine-polymer compatibility is evaluated, a series of tests are performed to select the potential candidates for a field application including filterability, molecular weight, chemical degradation, viscosity and density, screen factor, static adsorption, and thermal stability at reservoir temperature, as described below.

3.4.1 Polymer filterability

The filter ratio test is important to ensure that a polymer solution is free of aggregates which could lead to formation plugging [2, 11]. Filterability of polymers can be affected by the quality of the injection water, fluid incompatibilities, bacterial contamination, polymer chemistry, and stability. The best practice is to test both in formation and injection brines.

It is recommended to follow the standard protocols as reported in the literature [22, 23, 25–30], where filter ratios are evaluated at two polymer concentrations, usually 1000 and 2000 ppm.

A given volume (V_1) of the diluted polymer solution is introduced into a high-pressure filter apparatus loaded with a 0.1–5 μm polycarbonate filter. A pressure drop of 15 psi is applied and maintained during the test using nitrogen or argon gas making sure the flow rate remains constant to avoid the formation of aggregates. The time required to filter specific volumes of polymer solution (V_2, V_3) is recorded. The test is concluded once the total V_1 solution volume has been filtered. The filter ratio (FR) is calculated using Eq. (1).

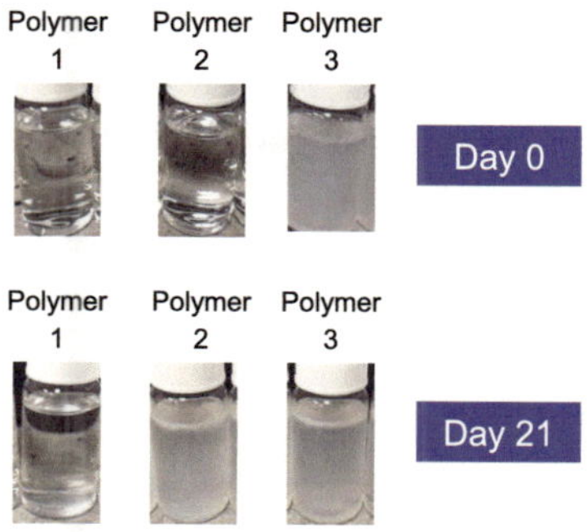

	Property	Polymer 1	Polymer 2	Polymer 3
Day 0	Viscosity (cP)	12.50	11.36	11.74
	Appearance	Clear	Clear	Cloudy
Day 21	Viscosity (cP)	12.46	11.23	11.41
	Appearance	Clear	Cloudy	Cloudy
Change in Viscosity (%)		0.32	1.14	2.81
Compatibility Result		C	NC	NC

Figure 2.
Example of compatible and non-compatible polymer-brine solutions.

$$FR = \frac{V_1 - V_2}{V_2 - V_3} \quad (1)$$

After the test, the filter is carefully inspected to see if any remaining polymer micro-gel is present due to improper hydration. A polymer solution with $FR < 1.2$ is considered acceptable for a rheological test [31]. The test is repeated using a 2000-ppm concentration for polymers with a $FR < 1.2$.

3.4.2 Polymer molecular weight

Even though the vendors provide this information with their products, it is a good practice to validate it (if possible). Several methods are widely used to evaluate MW, including Size Exclusion Chromatography [32], Gel Permeation Chromatography [33], Laser Desorption/Ionization Time-of-flight Mass Spectrometry [34], and Nuclear Magnetic Resonance [35]. Field flow fractionation [36] is another technique for MW measurement that is gaining popularity in the polymer user community. Calibration with well-characterized samples is required for accurate MW determination.

3.4.3 Chemical degradation

Polymers can be broken down into monomers in a process known as hydrolysis, a reaction done in the presence of water. Chemical degradation is evaluated through the degree of hydrolysis. Accepted values for polymer degree of hydrolysis are in the range of 15% and 33% [28, 37, 38]. Optimum values for polymer hydrolysis degree are in the range 25 +/− 5% [2, 11].

Quantification of the degree of hydrolysis for a polymer is an important step both before and after a polymer flood. It is important to consider that the degree and rate of hydrolysis increase with temperature and is a linear function of pH in non-neutral-pH solutions. For hydrolysis degrees greater than 30% and in the presence of calcium cations, PAM polymers precipitate with additional viscosity losses [38, 39].

The degree of hydrolysis can be evaluated by C13NMR spectroscopy and C, N elemental analysis [24] or by colloid titration based on the stoichiometric combination of positive and negative colloids where the endpoint is decided by indicators [40]. Colloidal titration allows for simple and accurate verification of a polymer's degree of hydrolysis. The polymer in solution can be precipitated, concentrated, and analyzed just like a dry polymer powder to assess if the polymer's degree of hydrolysis was altered during the flood.

Polymers can also be chemically degraded by redox reactions in the presence of contaminants, such as oxygen or iron, resulting in the formation of free radicals, that impact the polymer properties. In some cases, these unwanted effects can be reduced by using radical scavengers, chelating agents, or oxygen scavengers [5, 15, 41].

3.4.4 Viscosity, screen factor, and density of polymer solutions

It is important to evaluate the polymer viscosity and density as a function of its concentration and temperature at the optimal shear rate. The measurements should be performed using high precision instruments, such as low shear rheometers equipped with either ultra-low adapters or small sample adapters for viscosity, and high-temperature densitometers. The protocol associated with the viscosity and density measurement depends on the type of instrument used [11, 22].

It is recommended to measure viscosity at shear rates ranging from 0.1 to 500 s^{-1} for concentrations in the range of 0–5000 ppm. Density should be measured at reservoir temperature. The effect of brine salinity and pH should be assessed to quantify the decrease in viscosity with a salinity increase. Viscosity measurements are also used to evaluate the mechanical (shear) degradation of the polymer by shearing the polymer solutions at high rpm for different time intervals at a constant temperature. Viscosity monitoring at each interval allows assessing the mechanical degradation [3, 42].

The screen factor (*SF*), defined as the ratio of the flowing time of the polymer solution through the screen viscometer to the flowing time of the same volume of solvent, allows to measure the viscoelastic response of the polymer determining its ability to sustain sudden elongational deformation. It is considered as an optional test since it is more sensitive than the intrinsic viscosity to differences in molecular weight and molecular weight distribution.

As stated by Lim et al. [43], a major disadvantage of the intrinsic-viscosity measurement is the associated time-consuming experiments. This lengthy process must be carried out if meaningful results are to be obtained for high-molecular-weight polymers. *SF* measurements offer a simple, rapid characterization method for dilute polymer solutions that correlate well with end-use performance.

3.4.5 Static adsorption of polymer solutions

A key parameter in the application of polymers for chemical flood is the loss of polymer due to adsorption to the formation rock [3, 44, 45], and more in general, its retention during flow in the porous medium that can prevent proper oil displacement impacting the hydrocarbon recovery.

Since polymers are characterized by having high molecular weights and long chains typically containing many polar groups, they will tend to attach to the available polar points on the rock surface [46]. The adsorption can be severe for situations where large surface areas are available to the polymer. It is irreversible and can result in lower oil recovery [38, 47].

There are large differences (10–30%) in the way the adsorption of polymers is measured, either by static adsorption test or by the dynamic retention in a core or pack as shown by Lakatos et al. [48]. The differences are the result of changes in the specific surface area and the accessibility of certain regions of the pore space. Typically, the adsorption determined by the bulk static method is greater than the one determined by dynamic flow conditions [48].

The static adsorption test is done on crushed formation cores resulting in much higher adsorption levels, as more surface area is accessible to the polymer. The maximum accepted value for static adsorption is around 200 μg/g [47, 48].

A practical method used to quantify polymer adsorption is the approach suggested by Chiappa et al. [49], where for a given volume (V) of solution the polymer concentration is measured before (C_o) and after (C_e) exposure to formation rock. Polymer adsorption is calculated by dividing the loss of mass from the solution by the weight of the exposed rock (W).

$$q_{\text{abs}} = \frac{V(C_o - C_e)}{W} \tag{2}$$

Some criticisms of this method have been reported including—(a) it relies heavily on only two polymer concentration measurements, so errors in those measurements have a substantial impact on the calculated adsorption value [50]. (b) As the rock is pulverized, the surface area and minerals are exposed, this condition is not available during dynamic experiments (polymer solutions flowing through the porous rock) [38, 39, 50]. (c) The method cannot assess mechanically entrapped polymer [50, 51].

Polymer adsorption is highly related to the rock mineralogy and the presence of clays. In general, it is dominated by nanoclay and nanosilica [52]. Higher adsorption is observed in carbonates than in sandstones.

3.4.6 Thermal and biological stability of polymer solutions

The assessment of the long-term thermal stability of the polymer solution should be done with the polymer solution diluted to its target viscosity in oxygen-free environments (< 10 ppb) at the reservoir target temperature and in sealed glass ampoules. Details on the experimental setup are given by Araujo and Araujo [42].

The primary purpose of this test is to verify that the polymer remains stable in brine for 90 days. Stability in this context means the polymer solution remains clear and its viscosity does not decrease drastically during the aging period. Accepted criteria can be stated as follow: (i) Clear single-phase solution at both ambient and reservoir temperature observed; and (ii) viscosity changes <5% over the 90-day period, as shown in **Figure 3**.

Thermal stability is expressed as the percentage of the viscosity retained after a chosen period of exposure to a higher temperature.

The biological stability of a polymer solution can be tested in a similar way as the thermal stability by measuring the viscosity loss of the solution as a function of time [53]. An alternative method is to observe bacteria growth using light scattering and

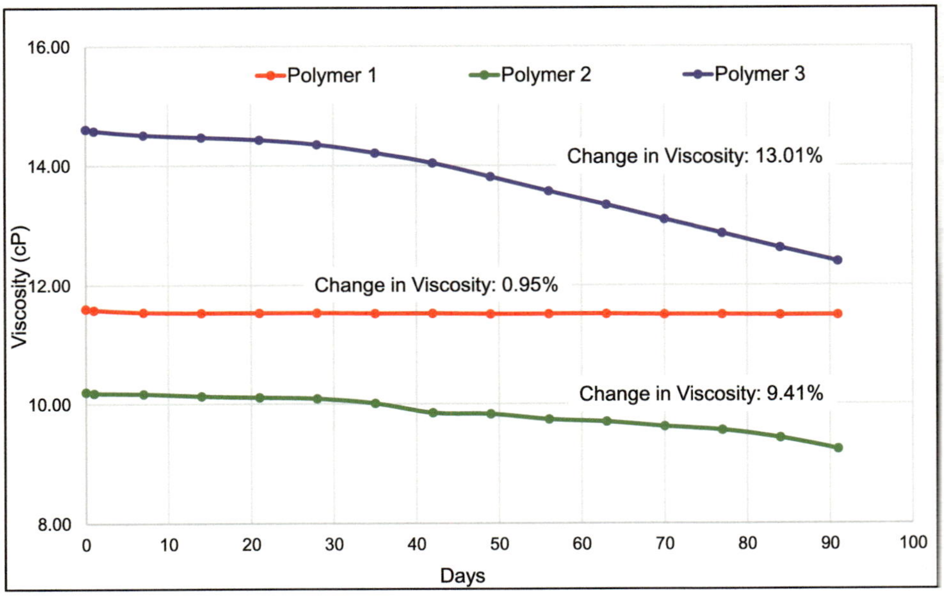

Figure 3.
*Example of thermal stability results for the polymers shown in **Figure 2**.*

other methods or to evaluate the bacteria metabolism [53]. Testing should also include the effectiveness of biocide.

3.5 Evaluation of the polymer performance

3.5.1 Rock and fluids preparation

To evaluate the performance of the preselected polymers, the first step is to select rock samples for testing and make sure reservoir fluids are available or could be reproduced in the laboratory. It is recommended to have representative rock samples based on rock and fluid characterization best practices [3, 11, 42, 54]. This is especially important since lithology differences can impact polymer performance.

Representative reservoir fluids should be used in all lab testing. In practice, the use of "live oil" could be limited by the quality and availability of reservoir oil samples. If this happens either bottom-hole samples or dead crude oil can be used or recombined with gas based on PVT data to properly mimic the *in situ* conditions [42]. After deciding which oil to use, it is important to determine relevant crude properties including viscosity, density, total acid number (TAN), SARA, and sand & water content (BSW).

All reservoir oil (whether live, recombined, or dead) must be filtered and treated to remove any contaminants (such as sand particles and water droplets) prior to making any measurements. For the filtration, the filter size needs to be considered especially for low-permeability samples. A good understanding of reservoir lithology provides steer on the filter selection. Standard filter sizes are 10, 3, and 0.45 μm. If the average mean pore throat size is smaller than these filter sizes, the selected core plug sample will be acting as a further filter during testing, with the potential risk of plugging.

A suitable oil sample can be prepared from dead oil and PVT data. In the process, the dead oil is mixed with a solvent to reduce the viscosity to its target value. The following procedure is recommended for its preparation: (i) Heat the dead oil in the original container (or cylinder) to reservoir temperature; (ii) shake the container vigorously to homogenize the oil and pour it into a beaker; (iii) stir the oil using overhead stirrer and add solvent (toluene), stir for a minimum of 3 hours; (iv) transfer the mixture into a fluid accumulator, and allow it to equilibrate in an oven in a vertical position for a half-day, then check the viscosity; (v) if the viscosity is too low, continue stirring to allow the solvent to evaporate, if the viscosity is too high, add solvent to the mixture. When working at a high temperature (that could lead to solvent evaporation), it is important to annotate the initial and final volumes to ensure there is solvent present at the end of the mixing step.

The laboratory brine is typically prepared using the reservoir formation and produced waters chemistry. Once prepared the brine properties are determined using the standard protocols. Recommended properties include evaluation of density, total dissolved solids (TDS), turbidity, pH, and conductivity. If the brine is not used immediately, it should be handled and stored following the procedure described by McPhee et al. [54].

Regarding the rock samples, it is important to use samples with the appropriate lithology and to consider the formation heterogeneity to make sure the selected samples are representative. For complex lithologies, such as turbidites and carbonates, it is recommended to use multiple samples rather than focusing on just one. Once a decision is made, the core samples (typically plugs or full core sections) are cleaned to remove any residual fluids and/or solids present. This is generally done via flushing, flowing, or solvent contact. The literature recommends using at least two of these methods for an effective cleaning [54]. Solvent selection is done based on lithology [54, 55]. Once the samples are clean, they are dried using an appropriate method, based again on the lithology, with special attention to mineral composition (frequently the drying is done under vacuum, or in a humidity-controlled environment). Drying is continued until no changes in sample weight are observed to make sure all interstitial water is removed.

It is very important to determine some rock properties prior to exposure of the rock to the polymer solution. It is recommended to evaluate the porosity, permeability, pore size distribution, wettability, and to have a mineralogy analysis. It is recommended to use core samples of at least seven in length since end effects can add uncertainty to the flow properties.

To closely relate to the actual reservoir conditions the wettability of the core samples should be restored via aging [54, 56]. This is done by exposing the clean rock sample to the reservoir fluids at the expected reservoir conditions. This step is time-consuming and thus not performed by many laboratories; however, it is very important for the understanding of the test results. Additional details are given in Ref. [54].

3.5.2 Coreflood testing

The primary purpose of the core flooding testing is to compare the selected polymers in terms of their propagation behavior through the porous formation including relative retention, permeability reduction, and impact on the pressure response to gather data for use in simulation models for EOR project design and field-testing planning.

During the coreflood test, the comparison of the polymer propagation is done via differential pressure measurement. It is recommended to measure the pressure gradient across various sections of the core and over the entire sample, and that the data show indications of no plugging or filtration effects of the injected polymer solution. The steady-state pressure differential should be stable and not change with an additional polymer injection.

Properties that are determined from core flooding tests include:

a. Pore volume (PV) and inaccessible pore volume (IAPV).

b. Mobility reduction or resistance factor (RF): defined as the apparent viscosity of the polymer solution during its flow through the porous formation. It is calculated as the ratio of the pressure drop across the sample during polymer injection and the one measured while flowing brine under steady-state conditions.

c. Polymer retention: informs the extent of the polymer propagation through the rock. Many methods are available to measure polymer retention including static and dynamic measurements [22, 49, 51, 57]. It is important to keep in mind that several factors may affect polymer retention including polymer type (cationic, anionic, and amphoteric), the degree of hydrolysis, the polymer and brine concentrations used, brine salinity, oil saturation, wettability [42], rock lithology and properties like permeability and pore size distribution. For details refer to Manichand and Seright [51]. Retention values measured at field scale range from 7 to 150 μg of polymer/cc of bulk volume with an acceptable retention level around 20 μg/cm^3 [45]. It is recommended that polymer retention be less than 100 ± 25 μg/g of rock.

d. Residual resistance factor (RRF) is defined as the loss of permeability given as the ratio of mobility of initial injected water to the mobility of injected water behind the polymer solution. It could be due to several mechanisms, such as polymer adsorption, mechanical retention of polymers in the pore space, and any other condition like precipitation that could retain the polymer formation (in particular when in presence of divalent cations in the formation water). It is determined as the ratio of the pressure differential post polymer injection and with brine at the same conditions.

Figure 4 shows the steps for core flooding testing to simulate polymer floods. The first step is to prepare the fluids (synthetic formation brine and injection brines) and restore the oil at reservoir conditions of temperature and pressure. The polymer solution is prepared at the target concentration defined by the rheology tests. Once prepared, the solutions are filtered, and the viscosity is measured before and after filtration at a shear rate of 7.34 s^{-1}. Then the polymer solution, brine, and oil are let to equilibrate (in an oven) at the reservoir temperature and pressure.

Simultaneously with the fluid preparation, the rock samples are cleaned, characterized, and restored according to the methods described earlier. After wettability restoration, two coreflood tests are performed:

a. Single-Phase test to determine the porous volume (PV), inaccessible pore volume (*IAPV*), injectivity response, brine permeability, polymer retention, resistance factor (*RF*), residual resistance factor (*RRF*), and the effluent characteristics including viscosity, pH, and polymer concentration in the effluent.

b. Two-Phase test to evaluate the retention of the polymer in the presence of oil, additional recovery from the polymer injection, and to optimize the polymer slug size.

A summary of the best practices for running the coreflood is given by Araujo and Araujo [3, 11, 42]. Two twin/similar samples are required for the single- and two-phase coreflooding tests for a proper assessment of the polymer performance.

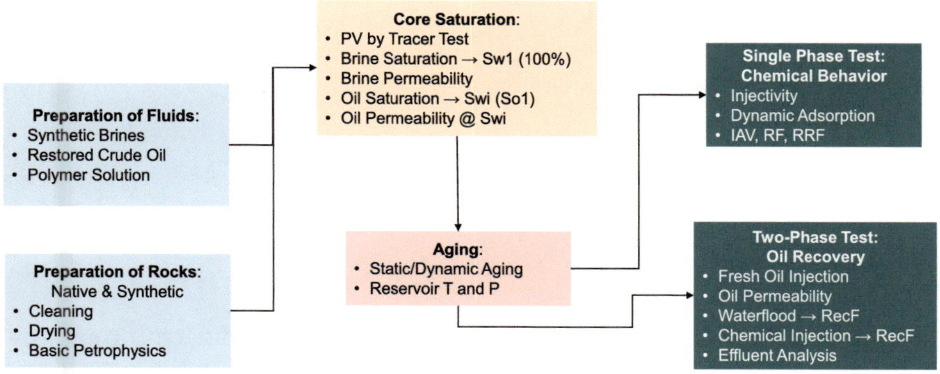

Figure 4.
General workflow for coreflooding testing using polymer.

3.5.2.1 Single phase test

Once the coreflood apparatus is setup and ready for testing with a new sample, the pressure is checked, and all lines are purged with brine before saturating the sample. The oven and fluid accumulators must be at reservoir temperature and the core holder at reservoir pressure. To measure the porous volume of the rock sample it is recommended to use a tracer test with a salt such as KCl [42]. After the test is complete, the system is flushed with brine to remove any residual KCl. The proposed single-phase test follows a modified Osterloh procedure [58]:

1. The core sample is fully saturated with formation brine.

2. Brine is injected into the core at different rates (1–32 ft./day) to determine brine permeability.

3. Polymer solution (at least 2PV) is injected into the core with a tracer using a flow rate selected from brine permeability testing (in most cases 2 ft./day).

4. Samples are collected in 4 ml increments, and the polymer and tracer (if used) concentrations are measured.

5. After the effluent concentrations for both the polymer and tracer reach the injected concentration levels, several PVs of brine are injected to displace the mobile polymer and the residual tracer.

6. The total brine injected is recorded and permeability reduction (resistance factor, *RF*) is calculated. The resistance factor is determined using Eq. (3), where ΔP Polymer (Q) is the pressure differential measured when injecting polymer at rate Q, and ΔP Brine (Q) is the pressure differential measured when injecting a brine at the same rate Q. It is a good practice to check that the resistance factor calculated for the entire core is similar to the one for sections of the core.

$$RF = \frac{(k/\mu)\text{brine}}{(k/\mu)\text{polymer}} = \frac{\Delta P \text{ Polymer } (Q)}{\Delta P \text{ Brine } (Q)} \tag{3}$$

7. After calculating *RF*, a second polymer flood is performed to estimate IAPV and quantify the amount of polymer retained in the core sample. It is important to keep in mind that polymer adsorption during the second polymer flood is very small, thus the polymer can be used as a tracer.

IAPV is determined as the difference in the areas under the polymer breakout curve and the tracer breakout curve using Eq. (4) [51, 57, 58].

$$IAPV = \sum\left[\left(\frac{C_p}{C_{pi}} x\Delta PV\right) - \left(\frac{C_t}{C_{ti}} x\Delta PV\right)\right] \tag{4}$$

where C_p is the polymer concentration in the effluent, C_t is the tracer concentration in the effluent, C_{pi} is the injected polymer concentration, C_{ti} is the injected tracer concentration, *PV* is the pore volume equivalent and ΔPV is the pore-volume increment.

8. The residual resistance factor (*RRF*) is determined using Eq. (5) [57, 58].

$$RRF = \frac{\Delta P \text{ water after Polymer Flooding}(Q)}{\Delta P \text{ water before Polymer Flooding}(Q)} \tag{5}$$

RRF shows how much reduction of permeability has occurred after polymer flooding in rock due to polymer molecules adsorption.

Polymer concentration can be measured by ICP-OES, the starch iodide method [59], and from viscosity via a calibration curve [60].

The experiment is repeated using another polymer solution (higher concentration) to determine *RF*, and at the end of the test, after injection of multiple PVs of brine, the *RRF* is calculated. Viscosity measurements done using an inline capillary viscometer can also be used to determine *RF* and *RRF* through viscosity measurements during the polymer injection [60]. Using the *RF* versus polymer throughput curves the potential plugging during the flooding can be estimated [57].

The injectivity is evaluated during the polymer flow by observing the trend of the pressure drop across the core. Since injectivity is one of the polymer selection criteria good candidates are formulations with associated low-pressure drop. Magnitude of the pressure differential across the sample varies according to lithology, with higher values for tight formations [42]. It is important to keep in mind that polymer mobility reduction could be observed at low injection rates attributed to polymer shear thinning behavior due to accumulation in the core.

Seright et al. [57] give a good discussion on the range of typical values of RF and RRF comparing scenarios with varying conditions (lithology, temperature, salinity, etc.).

3.5.2.2 Two-phase test

Using the second twin sample, the porous volume and the brine permeability are first determined. The oil permeability is determined next, which is done by injecting at least 2 PV of oil for at least three to five flow rates typically in the range of 1–64 ft./day. At those conditions, the sample is then aged for about 40 days to allow wettability restoration.

Once the sample is restored, a batch of oil is injected and the initial oil saturation (S_{oi}) is calculated by mass balance using Eq. (6).

$$S_{oi} = \frac{V_w}{PV} \tag{6}$$

where V_w is the volume of the produced water, and *PV* is the pore volume of the core sample. The irreducible water saturation (S_{wr}) is calculated using Eq. (7).

$$S_{wr} = 1 - S_{oi} \tag{7}$$

Effective oil permeability is calculated using Darcy's equation with the measured pressure drop and the volumetric flow rate, and the end-point oil relative permeability as the ratio of the effective oil permeability (k_o) to the brine permeability (k_b) as shown in Eq. (8).

$$k_{ro}^{o} = \frac{k_o}{k_b} \tag{8}$$

Once these properties are evaluated the flooding sequence starts with a waterflood for at least 3 PV making sure reaching steady-state conditions. The objective of this step is to displace and produce the mobile oil and determine the oil saturation after waterflood. The oil saturation after the waterflood step is referred to as the remaining oil saturation and it is not necessarily the residual oil saturation. The volume of oil produced during the waterflood is recorded and used to calculate the oil saturation at the end of the waterflood. For that purpose, a fractional effluent collector is used to gather effluents in glass tubes to calculate the oil recovery.

The remaining oil saturation after waterflooding S_{orw} is calculated using Eq. (9).

$$S_{orw} = \frac{(V_w - V_o)}{PV} \tag{9}$$

where V_w is the volume of the produced water with oil flood, V_o is the volume of oil produced with waterflood and PV is the pore volume of the core.

Effective water permeability is calculated using Darcy's equation and the end-point water relative permeability is calculated as the ratio of the effective water permeability (k_w) to the brine permeability (k_b) in the presence of S_{or} as per Eq. (10)

$$k_{rw} = \frac{k_w}{k_b} \tag{10}$$

Once no more oil is produced, up to 4 PV of the selected polymer at the target concentration are injected at the selected flow rate (chosen from the rates used to determine the brine permeability). The following data should be collected: pressure (total and differential), volume fractions, and viscosity of the produced fluids (oil and brines) as a function of time. After ending the polymer injection, brine is injected until the concentration of the polymer in the produced stream is very low (replicating the post-flush step done in the field).

At the end of the coreflood testing, it is recommended to do an additional polymer injection into a representative restored sample with the actual field saturation conditions (reproducing the saturation history of the field) to observe the behavior of the selected polymer candidates. This experiment can provide good insights on polymer performance and data for the simulation (including pseudo relative permeability curves), and it is typically not included in most laboratory screening and testing programs.

4. Final recommendations

Laboratory screening is an essential step in the design of polymer flooding. It is critical that the testing program is completed following validated workflows to avoid cost overruns and time delays.

We recommend using all information about the reservoir where the polymer injection will be done, collecting all required data on fluids, and doing the testing at reservoir conditions, for a better understanding of the formation response to the injection. In particular, polymer rheological properties tend to be affected by the chemical structure and other parameters.

Special attention should be given to the preparation and handling of all solutions. Vendor guidance on polymer preparation should be followed to avoid hydration issues upon mixing. All fluids should be properly characterized and stored to avoid degradation and/or contamination.

Among the recommending testing and sense check of the results here are some points to pay attention to—(a) evaluate the impact of polymer salinity on the solution viscosity, (b) analyze polymer rheology as a function of temperature to check for potential degradation [3, 11], (c) evaluate polymer solubility when using high salinity brine (due to strong interactions there could be changes in the polymer structure), (d) complete filtration ratio tests early in the experimental program, (e) check for biodegradation even if using synthetic polymers. If biocide is needed, make sure to select one that is compatible with the polymer, (f) check for the presence of any dissolved oxygen and reduced iron ($Fe^{+2)}$ to prevent chemical degradation, (g) complete long term degradation tests even under time constraints to make sure the degree of hydrolysis does no change significantly with time and temperature [26, 28, 29], (h) if oxygen scavengers and/or biocides are used, make sure they do not compromise the polymer structure.

We recommend using rock samples representative of the formation lithology where the polymer will be injected, with attention to capture data on the impact of heterogeneity on polymer performance. The rock samples should be restored to the reservoir state as close as possible, and the fluids (oil, brine) to their reservoir conditions. Special attention is given to cases where clay is present [3].

Core flooding tests are recommended for use to understand polymer behavior and avoid discarding candidates based on the results of a single test. Analyze the experimental results holistically. Aim to understand the obtained values of RF and RRF. Look out for other effects that could result in certain trends like higher values of RF or RRF due to gel-type effects in aqueous solutions.

We proposed an integrated workflow with reference to the recommended testing and Quality Check steps for a successful laboratory polymer screening program. We also recommended adding an additional step of injecting the selected polymer candidate at reservoir conditions in a formation sample reproducing the saturation history of the reservoir where the EOR process will be deployed. This last step can provide valuable data and insights for the process upscaling and field simulation.

Author details

Yani Araujo[1*] and Mariela Araujo[2]

1 SGS North America, The Woodlands, Texas, USA

2 Upstream Consultant, Houston, Texas, USA

*Address all correspondence to: yani.araujodeitriago@sgs.com

References

[1] Zhang J, Zhang F, Kang X, Li B. Development and Application of Chemical EOR Technologies in China Offshore Oil Fields. Rijeka: Intech; 2019. DOI: 10.5772/intechopen.88942

[2] Seright RS. Potential for polymer flooding reservoir with viscous oil. SPE Reservoir Evaluation and Engineering. 2010;**13**:730-740

[3] Pope GA. Overview of chemical EOR. In: Presented at the Casper EOR Workshop. Casper Wyoming, USA: University of Wyoming; 2007

[4] Gao CH. Scientific research and field applications of polymer flooding in heavy oil recovery. The Journal of Petroleum Exploration and Production Technology. 2011;**1**:65-70

[5] Thomas A, Gaillard N, Favéro C. Some key features to consider when studying acrylamide-based polymers for chemical enhanced oil recovery. Oil & gas Science and technology – Rev. FP Energies Nouvelles. 2012;**67**(6):887-902

[6] Ogezi O, Strobel J, Egbuniwe D, Bernd L. Operational aspects of a biopolymer flood in a mature oilfield. In: Presented at the SPE Improved Oil Recovery Symposium. Tulsa, Oklahoma, USA: OnePetro; 2014. DOI: 10.2118/169158-MS

[7] Skauge T, Djurhuus K, Hetland S, Spildo K, Skauge A. Offshore EOR implementation – LPS flooding. In: Presented at the 16th European Symposium on Improved Oil Recovery. Cambridge, Uk: University of Wyoming; 2011

[8] Chang HL. Polymer flooding technology - yesterday, today, and tomorrow. Journal of Petroleum Technology. 1978;**30**:1113-1128

[9] Standnes DC, Skjevrak I. Literature review of implemented polymer field projects. Journal of Petroleum Science and Engineering. 2014;**122**:761-775

[10] Selle OM, Fischer H, Standnes DC, Auflem IH, Lambertsen AM, Svela PE, et al. Offshore polymer/LPS injectivity test with focus on operational feasibility and near wellbore response in a Heidrun injector. In: Presented at the SPE Annual Technical Conference and Exhibition; 30 September – 2 October 2013. New Orleans, Louisiana, USA: SPE 166343; 2013. pp. 3103-3119

[11] Araujo YC, Araujo M. Polymers for application in high temperature and high salinity reservoirs: Critical review of properties and aspects to consider for laboratory screening. Revista Fuentes: El Reventón Energético. 2018;**16**(2):55-71

[12] Yongjun G, Zhang J, Zhang X, Hu J, Wang W, Yan L. Investigation and application of an associative polymer-surfactant binary system for a successful flooding pilot in a high-temperature, high-salinity, ordinary heavy oil reservoir. In: Presented at the SPE EOR Conference at Oil and Gas West Asia. Muscat, Oman: OnePetro; 2018. DOI: 10.2118/190411-MS

[13] Song K, Sun N, Pi Y. Laboratory study on EOR in offshore oilfields by variable concentrations polymer flooding. The Open Petroleum Engineering Journal. 2017;**10**:94-107. DOI: 10.2174/1874834101710010094

[14] Dupuis G, Antignard S, Giovannetti B, Gaillard N. A new thermally stable synthetic polymer for harsh conditions of Middle East reservoirs. Part I. thermal stability and injection in carbonate cores. In:

Presented at the Abu Dhabi International Petroleum Exhibition and Conference. Abu Dhabi, UAE: ADIPEC; 2017. DOI: 10.2118/188479-MS

[15] Thomas A, Braun O, Dutilleul J, Gathier F, Gaillard N, Leblanc T, et al. Design, characterization and implementation of emulsion-based polyacrylamides for chemical enhanced oil Recovery. In: Conference Proceedings of the 19th European Symposium on Improved Oil Recovery. Stavanger: EarthDoc; 2017. pp. 1-29. DOI: 10.3997/2214-4609.201700286

[16] Morel DC, Labastie A, Nahas E, Stephane J. Feasibility study for EOR by polymer injection in offshore fields. In: Presented at the International Petroleum Technology Conference. Dubai, UAE: IPTC; 2007

[17] Zhou W, Wang M, Sun G, Li C, Han P, Quan H, et al. Polyether-based thermoviscosifying polymers for enhanced oil recovery: Emulsion versus powder. Energy & Fuels. 2014; **34**(3):2824-2831. DOI: 10.1021/9b03928

[18] Jouenne S, Klimenko A, Levitt D. Tradeoffs between emulsion and powder polymers for EOR. In: Presented at the SPE Improved Oil Recovery Conference. Tulsa, Oklahoma, USA: OnePetro; 2016. DOI: 10.2118/179631-MS

[19] Rashidi M, Blokhus AM, Skauge A. Viscosity study of salt tolerant polymers. Journal of Applied Polymer Science. 2010;**117**(3):1551-1557

[20] Yerramilli SS, Zitha PL, Yerramilli RC. Novel insight into polymer injectivity for polymer flooding. In: Presented at the SPE European formation Damage Conference and Exhibition. Noordwijk, The Netherlands: OnePetro; 2013

[21] Akbari S, Mahmood S, Tan I, Ling O, Ghaedi H. Effect of aging, antioxidant and mono-and divalent ions at high temperature on the rheology of new polyacrylamide-based co-polymers. Polymers. 2017;**9**:480

[22] API. Recommended Practices for Evaluation of Polymers Used in Enhanced Oil Recovery Operations API RP 63. 1st ed. Washington: American Petroleum Institute; 1990. p. 87

[23] Levitt D, Pope GA. Selection and screening of polymers for enhanced-oil recovery. In: Presented at the SPE Symposium on Improved Oil Recovery. Tulsa, Oklahoma, USA: OnePetro; 2008. DOI: 10.2118/113845-MS

[24] Zurimendi JA, Guerrero SJ, Leon V. The determination of the degree of hydrolysis in poly(acrylamides): Simple methods using C13 NMR and elementary analysis. Polymer. 1984;**25**(9):1314-1316

[25] Dwarakanath V, Dean RM, Slaughter W, Alexis D, Espinosa D, Kim DH, et al. Permeability reduction due to use of liquid polymers and development of remediation options. In: Presented at the SPE Improved Oil Recovery Conference. Tulsa, Oklahoma, USA: OnePetro; 2016. DOI: 10.2118/179657-MS

[26] Ghosh P, Mohanty KK. Laboratory treatment of HPAM polymers for injection in low permeability carbonate reservoirs. Journal of Petroleum Science and Engineering. 2020;**185**:106574

[27] Kozicki W, Hsu CJ, Pasari SN. Evaluation of polymer adsorption-gel formation and slip in polymer solution flows. Chemical Engineering Communications. 1987;**59**(1–6):137-160

[28] Driver JW, Britton C, Hernandez R, Glushko D, Pope GA, Mojdeh D.

Conditioning polymer solutions for injection into tight reservoir rocks. In: Paper Presented at the SPE Improved Oil Recovery Conference. Tulsa, Oklahoma, USA: OnePetro; 2018. DOI: 10.2118/190217-MS

[29] Seright RS, Campbell A, Mozley P. Stability of partially hydrolyzed polyacrylamides at elevated temperatures in the absence of divalent cations. In: Presented at the SPE International Symposium on Oilfield Chemistry. The Woodlands,Texas, USA: SPE; 2009. DOI: 10.2118/121460-MS

[30] Abidin AZ, Puspasari T, Nugroho WA. Polymers for enhanced oil recovery technology. Procedia Chemistry. 2012;**4**:11-16

[31] Lee S, Kim DH, Huh C, Pope GA. Development of a comprehensive rheological property database for EOR polymers. In: Presented at the SPE Annual Technical Conference and Exhibition. New Orleans, Louisiana, USA: OnePetro; 2009. DOI: 10.2118/124798-MS

[32] Kulicke WM, Lange S, Heins S. Advantages of determining the molar mass distributions of water-soluble polymers and polyelectrolytes with FFFF—MALLS and SEC—MALLS. In: Chapter 9: Chromatography of Polymers. London, United Kingdom; Chapter 9: RSC Publishing; 1999. pp. 114-140

[33] Holzwarth G, Soni L, Schulz DN. Molecular weight distribution of water-soluble polymers: A new absolute method. Macromolecules. 1986;**19**(2): 422-426

[34] Yalcin T, Dai Y, Li L. Matrix-assisted laser desorption/ionization time-of-flight mass spectrometry for polymer analysis: Solvent effect in sample preparation. Journal of the American Society for Mass Spectrometry. 1998; **9**(12):1303-1310

[35] Izunobi IU, Higginbotham CL. Polymer molecular weight analysis by 1H NMR spectroscopy. Journal of Chemical Education. 2011;**88**(8): 1098-1104

[36] Messaud F, Sanderson R, Runyon J, Otte T, Pasch H, Williams S. An overview of field-flow fractionation techniques and their applications in the separation and characterization of polymers. Progress in Polymer Science. 2009;**34**:351-368

[37] Hirasaki GJ, Pope GA. Analysis of factors influencing mobility and adsorption in the flow of polymer solution through porous media. SPE Journal. 1974;**14**:337-346

[38] Green DW, Willhite GP. Enhanced Oil Recovery, Textbook Series. Vol. 6. Kuala Lumpur: SPE; 1998. pp. 100-185

[39] Sheng JJ. Modern Chemical Enhanced Oil Recovery: Theory and Practice. 1st ed. Gulf Professional, Elsevier Inc: Burlington; 2011. p. 647

[40] Terayama H. Method of colloid titration (a new titration between polymer ions). Journal of Polymer. 1952; **8**(2):243-225

[41] Gaillard N, Sanders D, Favero C. Improved oil recovery using thermally, and chemically protected compositions based on co-and ter-polymers containing acrylamide. In: Proceedings of the SPE Improved Oil Recovery Symposium. Tulsa, OK, USA: SPE; 2020. p. 24028

[42] Araujo YC, Araujo M. Polymers for EOR in offshore reservoirs: Recommended practices for laboratory screening. In: Presented at the Offshore

Technology Conference; 6-9 may 2019; OTC Houston, USA. DOI: 10.4043/29260-MS

[43] Lim T, Uhl JT, Prud'homme RK. The interpretation of screen-factor measurements. SPE Reservoir Engineering. OnePetro. 1986;**1**(03): 272-276. DOI: 10.2118/12285-PA

[44] Theng BKG. Chapter 2: Polymer behavior at clay and solid surfacesDevelopments in Soil Science, Formation and Properties of Clay-Polymer Complexes. Amsterdam, The Netherlands: Elsevier Science B.V.; 1979

[45] Sorbie KS. Polymer-Improved Oil Recovery. 1st ed. Dordrecht: Springer Science; 2013. p. 359. DOI: 10.1007/978-94-011-3044-8

[46] Ekanem EM, Rücker M, Yesufu-Rufai S, Spurin C, Ooi N, Georgiadis A, et al. Novel adsorption mechanisms identified for polymer retention in carbonate rocks. JCIS Open. 2021;**4**:100026

[47] Lee JJ, Fuller GG. Adsorption and desorption of flexible polymer chains in flowing systems. Journal of Colloid and Interface Science. 1985;**103**:569-577

[48] Lakatos I, Lakatos-Szabó J, Tóth J. Factors influencing polyacrylamide adsorption in porous media and their effect on flow behavior. In: Shah DO, editor. Surface Phenomena in Enhanced Oil Recovery. Boston, USA: Springer; 1981

[49] Chiappa L, Mennella A, Lockhart TP, Burrafato G. Polymer adsorption at the brine/rock interface: The role of electrostatic interactions and wettability. Journal of Petroleum Science and Engineering. 1999;**24**:113-122

[50] Zaitoun A, Kohler N. Two-Phase Flow through porous media: effect of an adsorbed polymer layer. In: Presented at the SPE Annual Technical Conference and Exhibition. Houston, Texas, USA: OnePetro; 1988. DOI: 10.2118/18085-MS

[51] Manichand RM, Seright RS. Field vs. laboratory polymer retention values for a polymer flood in the Tambaredjo field. In: Presented at the SPE Improved Oil Recovery Symposium. Tulsa, Oklahoma, USA: OnePetro; 2014. DOI: 10.2118/169027-MS

[52] Cheraghian G, Khalili SS, Kamari M, Hemmati M, Masihi M, Bazgir S. Adsorption polymer on reservoir rock and role of the nanoparticles, clay, and SiO2. International Nano Letter. 2014;**4**:114

[53] Littmann W. Polymer Flooding. 1st ed. Amsterdam: Elsevier; 1988. pp. 18-32

[54] McPhee C, Reed J, Zubizarreta I. Core Analysis: Best Practices. 1st ed. Amsterdam: Elsevier; 2015. pp. 136-178

[55] API. Recommended Practices for Core Analysis, API RP40. 2nd ed. Washington: American Petroleum Institute; 1998. p. 236

[56] Araujo YC, Araujo M, Molinaris J. Best practices for laboratory evaluation of immiscible WAG. In: Presented at the SPE Improved Oil Recovery Symposium. Tulsa, Oklahoma, USA: OnePetro; 2018. DOI: 10.2118/190303-MS

[57] Seright RS, Seheult JM, Talashek T. Injectivity characteristics of EOR polymers. SPE Reservoir Evaluation and Engineering. 2008;**12**(5):783-792

[58] Osterloh WT, Law EJ. Polymer transport and rheological properties for polymer flooding in the North Sea captain field. In: Presented at the SPE/DOE Improved Oil Recovery Symposium. Tulsa, Oklahoma, USA: OnePetro; 1998. DOI: 10.2118/39694-MS

[59] Scoggins MW, Miller JW. Determination of water-soluble polymers containing primary amide groups using the starch-triiodide method. SPEJ. 1979;**19**(3):151-154

[60] Moradi H. Experimental Investigation of Polymer Flow through Water- and Oil-Wet Porous Media [Thesis]. Stavanger, Norway: University of Stavanger; 2011

Section 5

Microbial Flooding

Chapter 9

Microbial Enhanced Oil Recovery: An Overview and Case Studies

Neha Sharma, Meeta Lavania, Nimmi Singh and Banwari Lal

Abstract

After traditional oil recovery processes, large amount of residual oil are still trapped in the pore spaces of the complex capillary network of the reservoir. MEOR (microbial enhanced oil recovery), a promising tertiary oil recovery method involves the utilization of indigenous microbial species capable of producing various secondary metabolites which further enhances the recovery of oil through their plugging, viscosity or interfacial tension reduction mechanisms. The chapter represents the overview of MEOR, mechanism involved in the process and field trials. Furthermore, microbial based mechanisms are widely demonstrated. The chapter confirms the credibility of MEOR process towards the enhanced oil recovery.

Keywords: selective plugging, biogases, solvents, case studies

1. Introduction

Generally, after the primary (natural pressure) and secondary recoveries (gas injection, water flooding etc.) of crude oil, around 35–55% of oil is left behind in the reservoir that needs to be extracted through other improved or enhanced oil recovery techniques [1]. There are many IOR and EOR techniques currently in practice such as miscible gas injection, polymer flooding and thermal EOR methods [2]. The selection of recovery methods is strongly influenced by the country's economy. Therefore, the development of low-cost technologies that bring maximum oil reserves to production is a major topic of energy research today [2].

2. Microbial enhanced oil recovery (MEOR)

An economical approach for the recovery of unrecovered oil is MEOR. MEOR often refers to injecting live microorganisms containing essential nutrients into oil reservoirs through injection wells. Under favorable environmental conditions in the reservoir, the infused microorganisms grow exponentially within the reservoirs and develop a variety of metabolites that play a crucial role in the mobilization of unrecovered oil that leads to enhanced oil recovery (**Figure 1**).

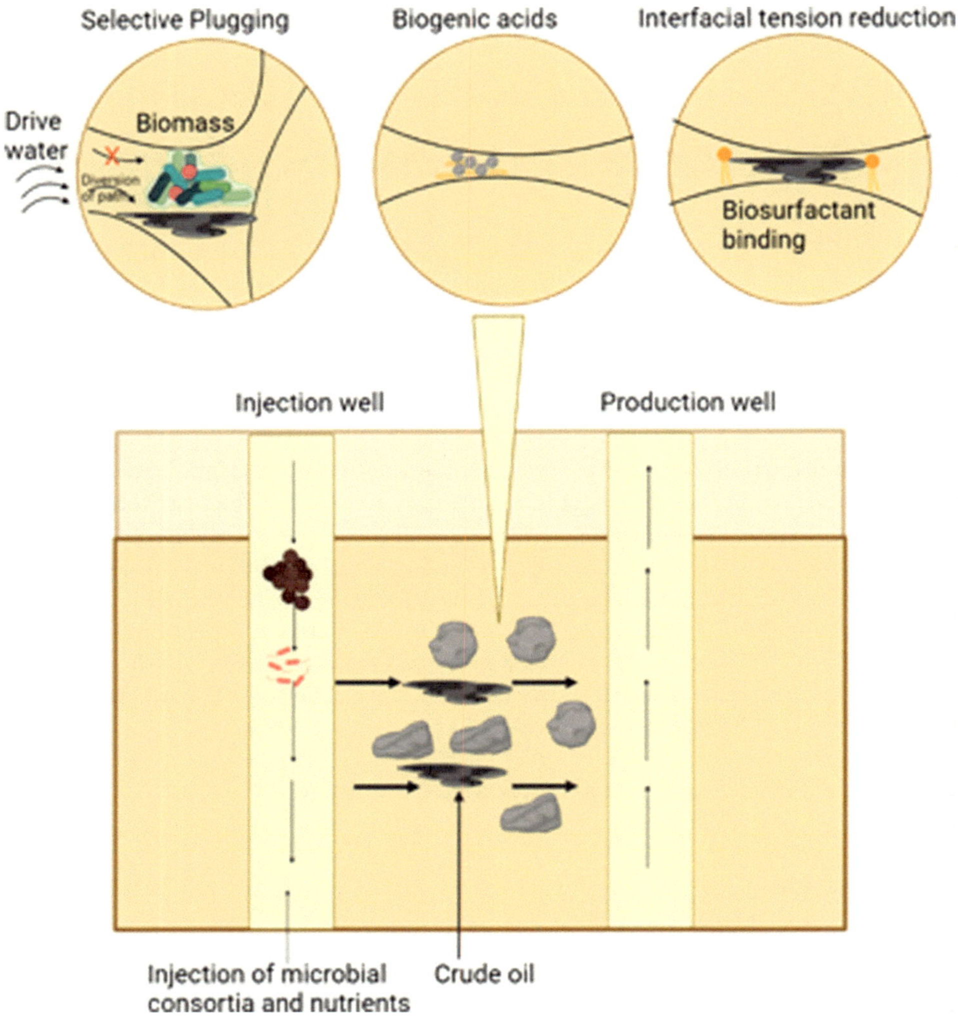

Figure 1.
Schematic representation of microbial enhanced oil recovery.

Various factors affect the growth of the microorganisms in oil reservoir *viz.* pressure, porosity, permeability, temperature, pH, dissolved solids, the salinity of the reservoir and availability of nutrients [3]. Therefore it becomes necessary to develop suitable microorganisms which can sustain the reservoir environment and can produce useful metabolites for recovering the oil. MEOR is believed to be able to extract up to 50% of the unrecovered oil remaining in a reservoir after primary and secondary recovery processes have been exhausted [1].

3. Requisite for MEOR

Instead of the recent renewable energy source available in the market, the world is still dependent on crude oil and petroleum-based products. To establish a green economical process viable research has been going to overcome various drawbacks such as extraction of crude oil from the reservoirs, reducing the harmful effect of crude oil on the environment etc. [1].

There are various types of recovery processes involved in oil extraction as shown in **Figure 2**. Firstly, primary oil recovery where the well pressure allows the gushing of oil from the reservoir led to less than 30% recovery of oil [4]. Once, the natural pressure of drilled well goes down other enhanced recovery methods were incorporated such a process is called secondary oil recovery methods. Secondary methods involve an injection of water at the wellhead that pushes the oil from the reservoirs. Secondary process accounts for 30–60% oil recovery [5]. To extract the remaining oil from the reservoir, an enhanced oil recovery process also known as a tertiary process can be used which involves surfactants, polymers, solvents, emulsifiers, acids, and dispersants along with the secondary method [6].

Microbial enhanced oil recovery involves two distinct approaches: First, bio augmentation where the exponentially grown microorganisms were injected in the reservoir that is capable of surviving and producing metabolites under harsh conditions. Eventually, the microbial species were occupied the metabolic niche within the oil reservoir. The second approach is an *in-situ* simulation where the nutrients were only added in the well that allows the inhabitant of the well to grow and proliferate to improve the recovery. Before finalizing the approach that needs to be implemented, researchers tend to explain the microbial niche found in the reservoir whether the niche is detrimental or beneficial [7].

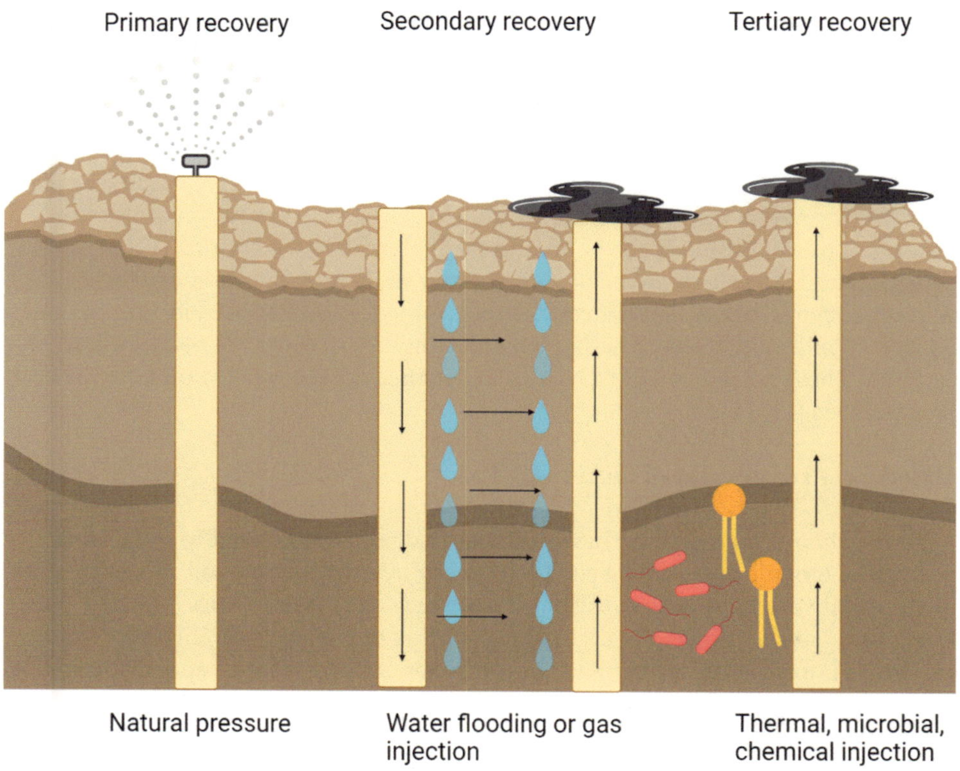

Figure 2.
Represents the type of recovery process for the extraction of oil.

4. Mechanisms involved in MEOR

The diverse types of mechanisms, metabolites are generated in the process of MEOR that make it complex and comprehensive. Insight knowledge of the MEOR mechanism is the source for determining the feasibility and efficiency of MEOR. In the MEOR process, nutrients are injected into the well in order to promote the proliferation of indigenous microbes. When the inorganic salts are added, the microorganism utilizes the crude oil as a carbon source and converts the complex form into the simple form. Conversely, when the exogenous carbon sources such as sucrose, glucose and molasses are added this led to the generation of more metabolites [1, 8].

The MEOR mechanism can be divided into two parts *viz.* utilizing biomass or biopolymer of microorganisms that selectively block the high permeable zones and facilitate the recovery of oil; another method is utilizing the solvents or biosurfactants that reduces the interfacial tension within the reservoir. If only the viscosity of crude oil is taken into account then recovery can be done in two ways, either reducing the oil-water interfacial tension or by reduction of viscosity through microorganisms or enzymes such as alkyl-monooxygenase/hydroxylase that degrade the heavy fraction of the oil [9].

4.1 Selective plugging

The major issue related to the recovery is the high porosity of the media found in the reservoirs. The oil saturates the media and accumulated in the inaccessible regions of the media such regions are called thief zones [10, 11]. The aim of the process is to release the entrapped oil from the thief zones, selective plugging involves the clogging of the media of high permeability that prevent the accumulation of oil and also after water flooding divert the water directly into the oil-rich zones that push the oil out of the reservoir. The biomass and biopolymers are used that are attached to the surface of the media where they proliferate. This led to the generation of biofilm and cluster that prevent the seepage of oil into the high permeable regions [12].

Biopolymer is the high molecular weight molecules metabolized by diverse microorganisms that contain hydroxyl groups which make it dipole, ion-dipole and hydrogen bonds with itself or with other substances to develop the network-like structure. These networks form a barrier to enhance the recovery of oil as reported in recent studies [13, 14].

4.2 Reduction of interfacial tension (IFT)

Biosurfactants are amphipathic molecules that contain both hydrophobic and hydrophilic moieties that can be produced *in situ* by microorganisms. Biosurfactants are effectively used in the recovery of oil from recalcitrant reservoirs [15]. The poor oil recovery is either due to the high viscosity of oil or low permeability of rock formation can be improved by utilizing the biosurfactants that have the capability to reduce the interfacial tension between the aqueous phase and oil saturation [16].

Biosurfactants are capable of reducing the capillary forces that halt the oil movement through the rock pores. Viscous forces are opposing force generated by the capillary forces that promote the flow in the reservoir. The parameter between the two forces is defined as a capillary number that accesses the chances of mobilization of residual oil within the reservoir. There is a direct correlation between the capillary number and mobilization of oil. Therefore, the biosurfactant can enhance the capillary number by reducing the IFT that facilitate the improved oil recovery [17].

4.3 Biogases, solvents and biogenic acids

MEOR mechanism involves the various metabolites such as solvents, gases and organic acids. Various thermophiles bacteria were reported that have the capability of synthesizing volatile fatty acid, biomass and gases that collectively help in the recovery of around 19% of oil in a core flood assay [13]. Different gases such as CO_2, H_2, CH_4 and other gases were produced during the fermentation of carbohydrates and other hydrocarbons. These gases help in pressurizing the crude oil and also reduce the viscosity of the oil.

Organic acids such as formic acid, acetic acid propionate that are low molecular weight were produced by the microorganisms during the fermentation process. These organic acids are capable of dissolving carbonate rocks that improves the reservoir permeability, whereas, gases and solvents can be reduced in viscosity by dissolving in crude oil [9].

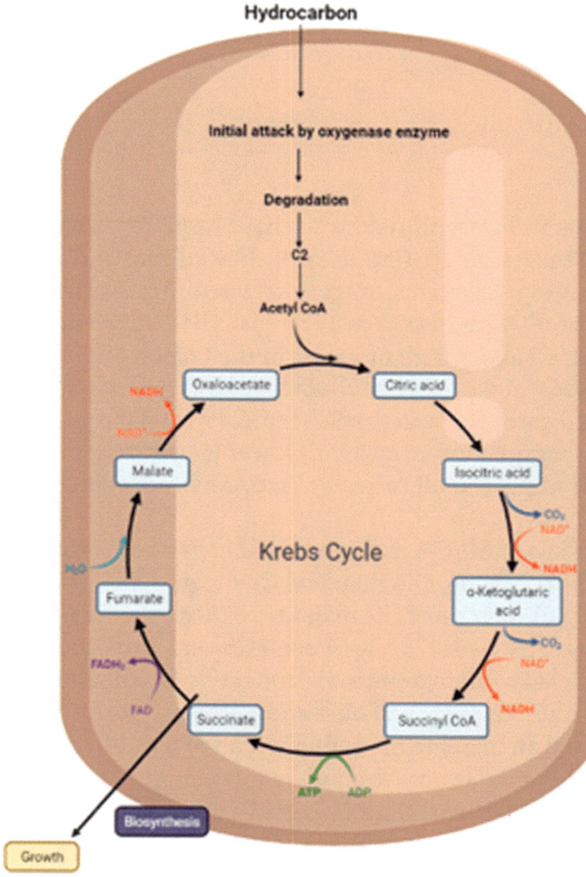

Figure 3.
Schematic diagram of steps involved in aerobic degradation of crude oil.

4.4 Biodegradation

One of the most attractive mechanisms is biodegradation in the MEOR process. In this process, the microorganisms utilizes the crude oil as a carbon source and convert the heavy fraction into the light components that fundamentally alters the viscosity, fluidity and properties of the crude oil thus, improving the oil recovery. Biodegradation is of two types such as aerobic and anaerobic biodegradation. In aerobic biodegradation initiated when the bacteria have taken up the crude oil that was oxidized by the oxygenase and peroxidases enzymes (**Figure 3**). The peripheral degradation pathway converts the organic components into various intermediates of central metabolism which led to the synthesis of the cell biomass from the precursor (Acetyl-CoA, succinate, and pyruvate) [18].

In general, the environment in a reservoir is anaerobic means no accessibility to oxygen, so various studies have reflected that hydrocarbon degradation was carried out by anaerobic microbes [19]. The in-depth investigation of the process proved that biochemical processes and genes involved belong to the anaerobic biodegradation [20].

On a laboratory scale, microbial-based enhanced oil recovery processes were also reported that showed diverse methanogens (*Methanothermobacter thermoautotrophicus*), bacterial species (*Thermoanaerobacter brockii*, *Thermoanaerobacter italicus*, *Thermoanaerobacter mathranii*, *Thermoanaerobacter thermocopriae*, *Clostridium* sp.) that significantly appeared an efficient model in core flood studies and can be implemented in the field trials [13, 16, 21].

5. Case studies

Worldwide, numerous MEOR field trials have been implemented that attain varying degrees of success. Statistical dataset showed that in field test more than 90% of MEOR trials represent encouraging affected [22]. United States was the first country to conduct MEOR field trial in 1954 with *Clostridium acetobutylicum*. The microbes were injected along with molasses in the Lisbon oil field of Arkansas which showed the generation of various metabolites such as gases, acid and biosurfactant [22, 23]. A field trial carried out in the Bebee field in Oklahoma showed the nine times lower biosurfactant concentration that was lower than the minimum concentration required for improved oil recovery, *Bacillus* strain was introduced in the field trail [24].

In China, field trials of MEOR had conducted in various oil fields such as Daqing, Shengli, Xinjiang, Jilin, Liaohe, Qinghai, and Changqing. In MEOR application, microbial wax removal was one of the major types used report suggested that a total of 11 fields have d in 1739 wells [25]. In Shengli Oilfield, huff and puff was carried out in 1640 wells, which led to an incremental production of 219,000 tons of crude oil, whereas in Daqing oilfield, microbial huff and puff were performed in 518 wells in 10 blocks that showed an increase in production up to 64,000 tons. In 2012, Daqing oilfield, microbial flooding was carried out in 45 injection wells that led to the total recovery of 56,837 tons of crude oil [26].

In Romania, various field trials were performed in the oil field that produced an average of 100 and 200% crude oil [22]. After the injection of hydrocarbon-degrading bacteria and anaerobic bacteria in Piedras Coloradas oil field in Argentina for 1 year, the average production of the wells was increased to 66% and the viscosity of the oil was also reduced effectively. Researchers were also tried to inject facultative anaerobic bacteria along with the nutrients for more than a year and there was a 20% incremental recovery of crude oil which accounts for 27,984 m^3 [27].

In Canada, endogenous microbial flooding was performed in the Saskatchewan oil field and results showed a 10% reduction in the water content in the first phase. In the second phase of the test (after 3 weeks), the production of crude oil was increased from 10.18 to 16.7 m^3/d [28]. In the Loco filed of Canada, a specially adapted strain of *Clostridium* was injected along with the nutrients showed promising results such as reduction of oil viscosity due to the production of carbon dioxide; it tends to improve the mobility and sweep efficiency [29].

In India, research on MEOR started in the nineties and since then MEOR processes involving in-situ stimulation and augmentation of microorganisms have been tested successfully with the production of enhanced production of oil. Oil & Natural Gas Corporation Ltd. (ONGC) had developed a thermophilic anaerobic bacterial consortium comprising Clostridium and Bacillus species and suitable for MEOR having reservoir temperature between 45°C and 65°C. Field trials were carried out in the Kosamba and Badarpur oil fields of ONGC with oil gain of 1150 m3 and 1200 m3 respectively. Further ONGC initiated research with The Energy and Resources Institute (TERI) and hyperthermophilic and halophilic anaerobic bacteria suitable for MEOR in oil reservoirs up to 90°C could be developed. The process has been patented in India, USA and Canada. A Joint Venture company ONGC-The Energy and Resources Institute Biotechnology Limited (OTBL) is implementing the MEOR process in huff and puff mode on a commercial scale. So far MEOR has been done in more than 125 oil wells mostly stripper wells with an average oil gain of about 300 m^3 per job/well. Further, research on the development of reservoir and oil specific bacterial consortia led to a yield of 2–8% additional oil recovery over and above waterflood recovery.

ONGC and TERI are in the process of developing the MEOR process for heavy oil reservoirs. Laboratory investigations and core/sand column flood tests are very impressive with a 13% recovery over waterflood recovery with 86% viscosity reduction after treatment of Becheraji oil. Another consortium could give a 13% recovery with a 42% reduction in oil viscosity of Lanwa oil. Field tests in five wells reported viscosity reduction of 17–25% in Becheraji wells and 11–18% in Lanwa wells. These fields were injected with strict anaerobic bacterial consortia habitant of the reservoirs. Statistically, 12 wells in four fields showed a threefold increase in crude oil production and efficient reduction in the water cut [16].

6. Advantages and limitations of the MEOR process

Numerous advantages associated with the MEOR process are cost-effective process as it involves the bacteria, nutrients and/or other natural products that are easily accessible, it is an economically attractive alternative, consume less energy as compared to the other EOR processes, the benefits of bacterial activity within the reservoir are amplified with time, whereas the opposite is true for other EOR technologies, the involvement of biodegradable products which make the process environmentally friendly [30]. A variety of metabolites such as acids, gas, solvents, surfactants, polymer etc. can be produced by bacterial consortia in the reservoir itself that can work simultaneously to recover the oil through their cumulative effect.

Another advantage including that the microbial processes can be simulated *in situ* within the reservoir, thus minimizing or eliminating the need to accommodate

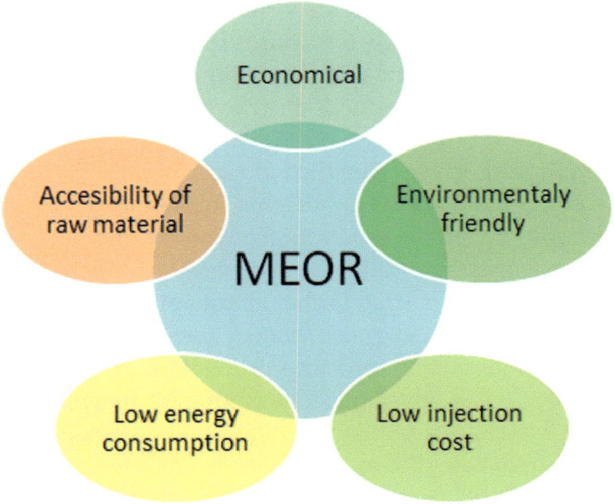

Figure 4.
Advantages of overall MEOR process compared to other recovery processes.

large storage facilities onsite/offshore (**Figure 4**). No extra infrastructure facility is required to set up for the execution of the MEOR process in the field, particularly for the *in-situ* simulation process.

Various limitations of MEOR processes are it is a complex process as specific bacterial activities are dependent on the condition of reservoirs. Majorly, MEOR field trials were conducted on the stripper wells reduces the microbial enhance recovery thus, impacted the oil recovery. MEOR is a slower process as it takes weeks or even months for proper outcomes. The production of microorganisms in the laboratory under desired reservoir conditions has been proven difficult.

7. Conclusion

Microbial enhanced oil recovery (MEOR) is advantageous over other recovery processes because of numerous reasons such as being environmentally friendly, economical; requiring fewer amounts of energy and operationally simple. In this chapter, the types of recovery, mechanisms of action and various fields' trials were reviewed that confirms the ultimate oil recovery using MEOR. To gain complete insight into microbial action assisted recovery, extensive field trials in different reservoir environment conditions are needed to generate datasets that will lead to the development of sustainable microbial systems and the defined mechanisms that are effective in the recovery of unrecovered crude oil which is left in the reservoir for a variety of reasons. Most of the trials have been done in huff and puff mode but large scale applications can better be done in flooding mode. The choice of *in-situ* bio-simulation and bio-augmentation approaches largely depends on the type of *in-situ* microflora present in the reservoir. The *in-situ* approach is more cost-effective with minimal or no effect on the ecosystem.

Author details

Neha Sharma, Meeta Lavania, Nimmi Singh and Banwari Lal*
Microbial Biotechnology, Environmental and Industrial Biotechnology Division,
The Energy and Resources Institute, Delhi, India

*Address all correspondence to: banwaril@teri.res.in

References

[1] Sen R. Biotechnology in petroleum recovery: The microbial EOR. Progress in Energy and Combustion Science. 2008;**34**(6):714-724

[2] Belyaev S et al. Use of microorganisms in the biotechnology for the enhancement of oil recovery. Microbiology. 2004;**73**(5):590-598

[3] Bachmann RT, Johnson AC, Edyvean RGJ. Biotechnology in the petroleum industry: An overview. International Biodeterioration & Biodegradation. 2014;**86**:225-237. DOI: 10.1016/j.ibiod.2013.09.011

[4] Gudina EJ et al. Biosurfactant-producing and oil-degrading Bacillus subtilis strains enhance oil recovery in laboratory sand-pack columns. Journal of Hazardous Materials. 2013;**261**:106-113

[5] Purwasena IA, Astuti DI, Syukron M, Amaniyah M, Sugai Y. Stability test of biosurfactant produced by *Bacillus licheniformis* DS1 using experimental design and its application for MEOR. Journal of Petroleum Science Engineering. 2019;**2019**:183

[6] Sałek K, Gutierrez T. Surface-active biopolymers from marine bacteria for potential biotechnological applications. AIMS Microbiology. 2016;**2**:92-107. DOI: 10.3934/microbiol.2016.2.92

[7] Da Silva MLB, Alvarez PJJ. Bioaugmentation. In: Handbook of Hydrocarbon and Lipid Microbiology. Springer; 2010. pp. 4531-4544

[8] Pereira JFB, Gudiña EJ, Costa R, Vitorino R, Teixeira JA, Coutinho JAP, et al. Optimization and characterization of biosurfactant production by Bacillus subtilis isolates towards microbial enhanced oil recovery applications. Fuel. 2013;**111**:259-268. DOI: 10.1016/j.fuel.2013.04.040

[9] Niu J, Liu Q, Lv J, Peng B. Review on microbial enhanced oil recovery: Mechanisms, modeling and field trials. Journal of Petroleum Science and Engineering. 2020;**192**:107350

[10] Hong E, Jeong MS, Lee KS. Optimization of nonisothermal selective plugging with a thermally active biopolymer. Journal of Petroleum Science and Engineering. 2019;**173**:434-446

[11] Okeke T, Lane R. Simulation and economic screening of improved-conformance oil recovery by polymer flooding and a thermally activated deep diverting gel. In: Proceedings of the SPE Western Regional Meeting. 2012

[12] Karimi M et al. Investigating wettability alteration during MEOR process, a micro/macro scale analysis. Colloids and Surfaces B: Biointerfaces. 2012;**95**:129-136

[13] Sharma N, Lavania M, Kukreti V, Rana DP, Lal B. Laboratory investigation of indigenous consortia TERIJ-188 for incremental oil recovery. Frontiers in Microbiology. 2018;**9**:2357

[14] Xu L et al. Synergy of microbial polysaccharides and branched-preformed particle gel on thickening and enhanced oil recovery. Chemical Engineering Science. 2019;**208**:115138

[15] Brown LR. Microbial enhanced oil recovery (MEOR). Current Opinion in Microbiology. 2010;**13**(3):316-320

[16] Sharma N, Lavania M, Kukreti V, Lal B. Instigation of

indigenous thermophilic bacterial consortia for enhanced oil recovery from high temperature oil reservoirs. PLoS One. 2020;**15**(5):e0229889

[17] Gray MR, Yeung A, Foght JM. Potential microbial enhanced oil recovery processes: A critical analysis. In: Proceedings of the SPE Annual Technical Conference and Exhibition; Denver, CO. 2008. pp. 3-27

[18] Sharma N, Lavania M, Lal B. Microbes and their secondary metabolites: Agents in bioremediation of hydrocarbon contaminated site. Achieves in Petroleum Environmental and Biotechnology. 2019;**4**(2):151. DOI: 10.29011/2574-7614.100051

[19] Lavania M, Cheema S, Lal B. Potential of viscosity reducing thermophillic anaerobic bacterial consortium TERIB#90 in upgrading heavy oil. Fuel. 2015;**144**:349-357. DOI: 10.1016/j.fuel.2014.12.003

[20] Callaghan AV, Wawrik B, Ní Chadhain SM, Young LY, Zylstra GJ. Anaerobic alkane-degrading strain AK-01 contains two alkylsuccinate synthase genes. Biochemical and Biophysical Research Communications. 2008;**366**(1):142-148

[21] Rathi R, Lavania M, Sawale M, Kukreti V, Kumar S, Lal B. Stimulation of an indigenous thermophillic anaerobic bacterial consortium for enhanced oil recovery. RSC Advances. 2015;**5**:88115-88124

[22] Safdel M, Anbaz MA, Daryasafar A, Jamialahmadi M. Microbial enhanced oil recovery, a critical review on worldwide implemented field trials in different countries. Renewable and Sustainable Energy Reviews. 2017;**74**:159-172

[23] Lazar I, Petrisor IG, Yen TE. Microbial enhanced oil recovery (MEOR). Petroleum Science and Technology. 2007;**25**(11-12):1353-1366

[24] Youssef N et al. In situ biosurfactant production by Bacillus strains injected into a limestone petroleum reservoir. Applied and Environmental Microbiology. 2007;**73**(4):1239-1247

[25] Wang W. Laboratory research and field trials of microbial oil recovery technique. Oil Drilling Products and Technology. 2012;**34**(1):107-113

[26] Gao C. Experiences of microbial enhanced oil recovery in Chinese oil fields. Journal of Petroleum Science and Engineering. 2018;**166**:55-62

[27] Strappa LA, De Lucia JP, Maure MA, Llopiz MLL. A Novel and Successful MEOR Pilot Project in a Strong Water-Drive Reservoir Vizcacheras Field. Argentina: Society of Petroleum Engineers, Tulsa, Oklahoma; 2004. p. 24

[28] Town K, Sheehy AJ, Govreau BR. MEOR success in southern Saskatchewan. SPE-158022-PA 13. 2010;(5):773-781

[29] Davidson S, Russell H. A MEOR pilot test in the Loco field. In: Proceedings of the Symposium on the Application of Microorganisms to Petroleum Technology. 1988

[30] Udipta Saikia BR, Vendhan E, Santosh KY, Siva SS. A brief review on the science, mechanism and environmental constraints of microbial enhanced oil recovery (MEOR). International Journal of ChemTech Research. 2013;**5**(3):1205-1212